# Reviews of Environmental Contamination and Toxicology

Editor
David M. Whitacre



VOLUME 207



Springer

ISBN 978-1-4614-2629-5 ISBN 978-1-4419-6406-9 (eBook)
DOI 10.1007/978-1-4419-6406-9
Springer New York Dordrecht Heidelberg London



Printed on acid-free paper

Springer is part of Springer Science+Business Media (www.springer.com)

# Reviews of Environmental Contamination and Toxicology

VOLUME 207

For further volumes:
http://www.springer.com/series/398

# Foreword

International concern in scientific, industrial, and governmental communities over traces of xenobiotics in foods and in both abiotic and biotic environments has justified the present triumvirate of specialized publications in this field: comprehensive reviews, rapidly published research papers and progress reports, and archival documentations. These three international publications are integrated and scheduled to provide the coherency essential for non-duplicative and current progress in a field as dynamic and complex as environmental contamination and toxicology. This series is reserved exclusively for the diversified literature on "toxic" chemicals in our food, our feeds, our homes, recreational and working surroundings, our domestic animals, our wildlife, and ourselves. Tremendous efforts worldwide have been mobilized to evaluate the nature, presence, magnitude, fate, and toxicology of the chemicals loosed upon the Earth. Among the sequelae of this broad new emphasis is an undeniable need for an articulated set of authoritative publications, where one can find the latest important world literature produced by these emerging areas of science together with documentation of pertinent ancillary legislation.

Research directors and legislative or administrative advisers do not have the time to scan the escalating number of technical publications that may contain articles important to current responsibility. Rather, these individuals need the background provided by detailed reviews and the assurance that the latest information is made available to them, all with minimal literature searching. Similarly, the scientist assigned or attracted to a new problem is required to glean all literature pertinent to the task, to publish new developments or important new experimental details quickly, to inform others of findings that might alter their own efforts, and eventually to publish all his/her supporting data and conclusions for archival purposes.

In the fields of environmental contamination and toxicology, the sum of these concerns and responsibilities is decisively addressed by the uniform, encompassing, and timely publication format of the Springer triumvirate:

*Reviews of Environmental Contamination and Toxicology* [Vols. 1–97 (1962–1986) as Residue Reviews] for detailed review articles concerned with any aspects of chemical contaminants, including pesticides, in the total environment with toxicological considerations and consequences.

*Bulletin of Environmental Contamination and Toxicology* (Vol. 1 in 1966) for rapid publication of short reports of significant advances and discoveries in the fields of air, soil, water, and food contamination and pollution as well as methodology and other disciplines concerned with the introduction, presence, and effects of toxicants in the total environment.

*Archives of Environmental Contamination and Toxicology* (Vol. 1 in 1973) for important complete articles emphasizing and describing original experimental or theoretical research work pertaining to the scientific aspects of chemical contaminants in the environment.

Manuscripts for Reviews and the Archives are in identical formats and are peer reviewed by scientists in the field for adequacy and value; manuscripts for the *Bulletin* are also reviewed, but are published by photo-offset from camera-ready copy to provide the latest results with minimum delay. The individual editors of these three publications comprise the joint Coordinating Board of Editors with referral within the board of manuscripts submitted to one publication but deemed by major emphasis or length more suitable for one of the others.

Coordinating Board of Editors

# Preface

The role of *Reviews* is to publish detailed scientific review articles on all aspects of environmental contamination and associated toxicological consequences. Such articles facilitate the often complex task of accessing and interpreting cogent scientific data within the confines of one or more closely related research fields.

In the nearly 50 years since *Reviews of Environmental Contamination and Toxicology* (formerly *Residue Reviews*) was first published, the number, scope, and complexity of environmental pollution incidents have grown unabated. During this entire period, the emphasis has been on publishing articles that address the presence and toxicity of environmental contaminants. New research is published each year on a myriad of environmental pollution issues facing people worldwide. This fact, and the routine discovery and reporting of new environmental contamination cases, creates an increasingly important function for *Reviews*.

The staggering volume of scientific literature demands remedy by which data can be synthesized and made available to readers in an abridged form. *Reviews* addresses this need and provides detailed reviews worldwide to key scientists and science or policy administrators, whether employed by government, universities, or the private sector.

There is a panoply of environmental issues and concerns on which many scientists have focused their research in past years. The scope of this list is quite broad, encompassing environmental events globally that affect marine and terrestrial ecosystems; biotic and abiotic environments; impacts on plants, humans, and wildlife; and pollutants, both chemical and radioactive; as well as the ravages of environmental disease in virtually all environmental media (soil, water, air). New or enhanced safety and environmental concerns have emerged in the last decade to be added to incidents covered by the media, studied by scientists, and addressed by governmental and private institutions. Among these are events so striking that they are creating a paradigm shift. Two in particular are at the center of ever-increasing media as well as scientific attention: bioterrorism and global warming. Unfortunately, these very worrisome issues are now superimposed on the already extensive list of ongoing environmental challenges.

The ultimate role of publishing scientific research is to enhance understanding of the environment in ways that allow the public to be better informed. The term "informed public" as used by Thomas Jefferson in the age of enlightenment conveyed the thought of soundness and good judgment. In the modern sense, being "well informed" has the narrower meaning of having access to sufficient information. Because the public still gets most of its information on science and technology from TV news and reports, the role for scientists as interpreters and brokers of scientific information to the public will grow rather than diminish. Environmentalism is the newest global political force, resulting in the emergence of multinational consortia to control pollution and the evolution of the environmental ethic. Will the new politics of the twenty-first century involve a consortium of technologists and environmentalists, or a progressive confrontation? These matters are of genuine concern to governmental agencies and legislative bodies around the world.

For those who make the decisions about how our planet is managed, there is an ongoing need for continual surveillance and intelligent controls to avoid endangering the environment, public health, and wildlife. Ensuring safety-in-use of the many chemicals involved in our highly industrialized culture is a dynamic challenge, for the old, established materials are continually being displaced by newly developed molecules more acceptable to federal and state regulatory agencies, public health officials, and environmentalists.

*Reviews* publishes synoptic articles designed to treat the presence, fate, and, if possible, the safety of xenobiotics in any segment of the environment. These reviews can be either general or specific, but properly lie in the domains of analytical chemistry and its methodology, biochemistry, human and animal medicine, legislation, pharmacology, physiology, toxicology, and regulation. Certain affairs in food technology concerned specifically with pesticide and other food-additive problems may also be appropriate. Because manuscripts are published in the order in which they are received in final form, it may seem that some important aspects have been neglected at times. However, these apparent omissions are recognized, and pertinent manuscripts are likely in preparation or planned. The field is so very large and the interests in it are so varied that the editor and the editorial board earnestly solicit authors and suggestions of underrepresented topics to make this international book series yet more useful and worthwhile.

Justification for the preparation of any review for this book series is that it deals with some aspect of the many real problems arising from the presence of foreign chemicals in our surroundings. Thus, manuscripts may encompass case studies from any country. Food additives, including pesticides, or their metabolites that may persist into human food and animal feeds are within this scope. Additionally, chemical contamination in any manner of air, water, soil, or plant or animal life is within these objectives and their purview.

Manuscripts are often contributed by invitation. However, nominations for new topics or topics in areas that are rapidly advancing are welcome. Preliminary communication with the editor is recommended before volunteered review manuscripts are submitted.

Summerfield, NC, USA David M. Whitacre

# Contents

# Contributors

**Rebecca Currie** The Dow Chemical Company, Midland, MI, USA, rjcurrie@dow.com

**Gary Klečka** The Dow Chemical Company, Midland, MI, USA, gmklecka@dow.com

**Dean Meyer** Centers for Disease Control and Prevention, National Center for Emerging and Zoonotic Infectious Diseases, 1600 Clifton Road, Atlanta, GA 30333, USA, lzz5@cdc.gov

**Carolyn Persoon** The University of Iowa, Iowa City, IA, USA, carolynpersoon@yahoo.com

**John F. Risher** Agency for Toxic Substances and Disease Registry, Division of Toxicology (F-32), Toxicology Information Branch, 1600 Clifton Road, Atlanta, GA 30333 USA, jrisher@cdc.gov

**G. Daniel Todd** Agency for Toxic Substances and Disease Registry, Division of Toxicology (F-32), Toxicology Information Branch, 1600 Clifton Road, Atlanta, GA 30333 USA, gtodd@cdc.gov

**Christie L. Zunker** Department of Neuroscience, Neuropsychiatric Research Institute, PO Box 1415, Fargo, ND 58103, USA, czunker@nrifargo.com

# Chemicals of Emerging Concern in the Great Lakes Basin: An Analysis of Environmental Exposures

**Gary Klečka, Carolyn Persoon, and Rebecca Currie**

## Contents



## 1 Introduction

Environmental analysis and monitoring have long been recognized as a means for assessing environmental quality. Within the Great Lakes watershed, the governments of the United States and Canada, together with collaborating agencies, have performed numerous surveys of environmental contaminants in the air, water, sediments, and biota. Environmental monitoring programs are necessary to develop

G. Klečka (✉)
The Dow Chemical Company, Midland, MI, USA
e-mail: gmklecka@dow.com

D.M. Whitacre (ed.), *Reviews of Environmental Contamination and Toxicology*,
Reviews of Environmental Contamination and Toxicology 207,
DOI 10.1007/978-1-4419-6406-9_1, © Springer Science+Business Media, LLC 2010

comprehensive descriptions of environmental quality, including at spatial and temporal scales, and to provide a sound basis for effective measures, strategies, and policies to address environmental problems (Calamari et al. 2000). While an important use of monitoring data is to inform environmental risk assessment, information gained from environmental measurements is also important for priority setting in regard to addressing the potential hazards of chemical contaminants.

Over the past 10 years, the emphasis on monitoring has shifted from the analysis of the so-called legacy pollutants to a wide array of new chemicals that have been discovered in the environment. These newer pollutants are lumped collectively into a group referred to as “chemicals of emerging concern.” While it has been known for over 20 years that compounds such as pesticides, detergents, personal care products, and pharmaceuticals enter the environment, the improvements in instrumentation and analytical methodology for detecting chemical substances in various environmental media (air, water, sediment, biota) have brought increased awareness and concern over the presence and potential risk that these chemicals may pose (Daughton 2001). Although thousands of chemicals are listed on chemical inventories in both the United States and Canada, very few are regulated as to their environmental release. The term “chemicals of emerging concern” increasingly is being used to characterize those chemicals used by society that are unregulated or inadequately regulated, and for which there is growing concern over the risk they may pose to the health of humans and ecosystems.

The topic of chemicals of emerging concern is not new to the International Joint Commission (IJC) and its advisory Boards. The Commission is an independent binational organization established by the governments of the United States and Canada under the Boundary Waters Treaty of 1909. Among the responsibilities of the Commission, under the Great Lakes Water Quality Agreement of 1978 and 1987 amendment, is the goal to restore and maintain the chemical, physical, and biological integrity of the Great Lakes Basin Ecosystem. Because the Great Lakes Water Quality Agreement focuses on a wide variety of water quality issues facing the Great Lakes Basin Ecosystem, the Commission created a priority-setting process to emphasize what it considers the most pressing issues. The Commission and its advisory bodies review and revise these priorities every 2 years as needed.

The topic of chemicals of emerging concern was addressed by the IJC Great Lakes Science Advisory Board with its *Expert Consultation on Emerging Issues of the Great Lakes in the 21st Century*, held February 5–7, 2003 at Wingspread, WI. Several papers in the IJC 2003–2005 Priorities Report dealt with this (chemicals of emerging concern) issue. Muir et al. (2006) summarized the various means for tracking, categorizing, and assessing chemicals in commerce and presented an overview of recent measurements of “new” chemicals in the Great Lakes. Walker (2006) addressed whether currently available tools, such as quantitative structure–activity relationships, can identify emerging pollutants that will threaten the Great Lakes ecosystem. Fox (2006) discussed the importance of monitoring programs in the context of meeting the requirements of the Great Lakes Water Quality Agreement.

In October 2007, the Commission began work on the 2007–2009 Nearshore Framework Priority. The purpose of this Priority is to assemble and report on the latest scientific, policy, and governance information on the nearshore of the Great Lakes so as to assess the binational implications of nearshore conditions and stressors. Nearshore problems are pressing and have significant social, economic, and environmental impacts. Current nearshore water quality is being adversely impacted by increased human population and problems arising from the existence of impervious surfaces and fertilizer use. Nearshore water quality is also influenced by land-based discharges from urban and agricultural sources, sediment resuspension, habitat loss and degradation, and atmospheric deposition, as well as by offshore waters. As the population increases, sewage discharges to receiving waters increase and impinge on water quality in the nearshore. Water quality in the nearshore is important to fish, aquatic birds, amphibians, and reptiles, since nearly all fish species spawn, have nursery grounds, and feed in the nearshore at some time in their development. The link between land-based activities and the nearshore has become recognized as the key challenge to protecting and restoring the chemical, physical, and biological integrity of the waters of the Great Lakes Basin Ecosystem.

Within the context of the 2007–2009 Nearshore Framework Priority, a multi-board work group was assembled to address the Priority on Chemicals of Emerging Concern. This group was charged with reviewing the scientific and policy aspects related to identification, impact, and management of chemicals of emerging concern in the Great Lakes. As a first step, a review of the status of current scientific knowledge on the occurrence of chemicals of emerging concern in the Great Lakes watershed was performed.

The objectives of this report, prepared for the multi-board work group, were to review and compile all peer-reviewed scientific studies and reports that had emerged since 1997; the scope of the review was on chemicals of emerging concern that could pose threats to water quality in the Great Lakes watershed. Emphasis was placed on chemicals discharged to the Great Lakes nearshore waters from wastewater treatment plants, as well as from other point and non-point sources of rural and urban pollution. The concentrations of chemicals in various environmental media disclosed by the review were assembled into a database, which was statistically analyzed to develop a quantitative understanding of current environmental exposures. To develop an initial assessment of their potential ecological significance, the concentrations found were compared with currently available regulatory standards, guidelines, or criteria.

## 2 Methods: Data Identification and Analysis

### *2.1 Identification and Critical Evaluation of Studies*

A literature search was conducted to identify recent environmental analysis studies (i.e., published between 1997 and 2008) which contained information on ecological exposures and the concentrations of chemicals of emerging concern in the Great Lakes Basin and watershed. The search terms were developed to capture as much information as possible on a wide variety of potential contaminants, in relevant environmental media (surface waters, sediment, biota, etc.). Excluded from consideration was information on concentrations reported in the atmosphere, precipitation, groundwater, and wastewater treatment effluents and biosolids. In addition to the literature search, input from experts in the field of environmental measurements and data from existing databases (Environment Canada, US Environmental Protection Agency, US Geological Survey, etc.) were sought as sources of information.

Each study that was found was critically evaluated for reliability using the review process described by Klečka et al. (2007); this process is generally based on internationally recognized guidance for the evaluation of measured data to be used in risk assessment (Canadian Environmental Protection Act (CEPA) 1997; European Commission (EUC) 2003a). All of the studies found were determined to be acceptable for use in this investigation.

## *2.2 Description of the Database*

A database was constructed from all acceptable data in Microsoft Excel and contained detailed information for each sample obtained in each study. Information on samples included in the database were as follows: location, source, date of sample collection, type of matrix, reported concentrations found, analytical methodology, method detection limits, quality assurance and quality control information, reliability rating for the study, references, and comments.

## *2.3 Statistical Treatment of Data*

Summary statistics were developed for the various datasets, including the number of data points, frequency of detection, average concentrations, standard deviation, minimum and maximum concentrations, and 5th and 95th percentile values, using the calculation functions in Excel. A considerable portion of the analytical values in the database were reported as being below the detection limit of the analytical method. For the statistical analysis, a variety of methods were considered for dealing with censored (i.e., non-detect) data. For this investigation, all values reported as "non-detectable" were set at one-half the method detection limit or minimum residue level (depending on the information available), since this is a long-established technique, which has been accepted by the regulatory community.

Much of available data in the database consisted of concentration values for various analytes in individual samples. However, in several studies, only summary statistics were reported for measured concentrations (i.e., information was provided for the number of samples, frequency of detection, minimum and maximum concentrations, etc.). When such studies were encountered, the corresponding authors were contacted when appropriate, and in many cases, the raw data were obtained. When only summary data were available (e.g., min–max), the maximum reported concentrations were used in the present analysis. When a dataset for a contaminant of interest consisted of a mixture of single-point and summary data, collections of related single-point data were initially analyzed to develop summary results, which were then combined in the statistical analysis of the remaining data.

# 3 Chemicals of Emerging Concern in the Great Lakes Basin

A total of 80 papers or reports, published between 1997 and 2008, were identified that contained environmental measurements for a variety of chemicals of emerging concern in the Great Lakes watershed. Although numerous papers were initially identified in the literature search, some studies were excluded because they were outside the cogent geographic boundaries or were published in the early 1990s. In addition, many other prospectively useful papers initially identified contained information on persistent organic pollutants or on other historical contaminants (e.g., DDT, polychlorinated biphenyls (PCBs).) that were considered to be outside the scope of this investigation. All of the papers relevant to this study were judged to be either reliable or very reliable. A summary of the assembled information on concentrations of chemicals of emerging concern in the Great Lakes Basin is shown in Table 1.

**Table 1** Summary of available information on concentrations of chemicals of emerging concern in the Great Lakes Basin

| Category | Number analytes | Data points by media | Sampling period |
|---|---|---|---|
| Current-use pesticides | 88 | Water = 12,471<br>Sediment = 296 | 1992–2006 |
| Pharmaceuticals | 60 | Water = 980<br>Sediment = 20<br>Biota = 21 | 1999–2006 |
| Organic wastewater contaminants; personal care products; steroids and hormones | 66 | Water = 267<br>Sediment = 69 | 1997–2006 |
| Alkylphenol ethoxylates | 24 | Water = 385<br>Sediment = 361<br>Biota = 163 | 1994–2002 |
| Synthetic musks | 9 | Water = 375<br>Sediment = 48<br>Biota = 42 | 1999–2006 |
| Perfluorinated surfactants | 31 | Water = 241<br>Biota = 400 | 1998–2003 |
| Polybrominated diphenyl ethers | 28 | Sediment = 45<br>Biota = 2,152 | 1979–2006 |
| Other flame retardants | 10 | Water = 32<br>Sediment = 119<br>Biota = 509 | 1994–2006 |
| Chlorinated paraffins | 10 | Water = 90<br>Sediment = 70<br>Biota = 455 | 1996–2004 |

From the 80 reviewed papers, a database was developed which contained a total of 19,611 data points, representing the results of the analysis of single samples or summary data (ranges, minimum–maximum). Concentrations in surface waters ($n = 14{,}841$) and biota ($n = 3{,}742$) represented the majority of the available data. Fewer values were reported for sediments ($n = 1{,}028$). Although the atmosphere is recognized as being an important medium for environmental transport of chemicals, and serves as a source for other environmental media, studies that focused on atmospheric concentrations (air, precipitation, etc.) were excluded from the present analysis.

The majority of the papers reviewed and used focused on a single category or group of chemicals, although one described the results of a nationwide reconnaissance program performed on the presence of organic wastewater contaminants in surface waters (Kolpin et al. 2002). In many of the papers, sampling locations were characterized as being downstream from municipal wastewater discharges, receiving waters for industrial facilities, areas susceptible to agricultural or urban contamination, or harbors or ports.

We divided the substances in the database into nine categories and will deal separately with each category. In each section below, we first provide a brief overview of the studies found that provided concentrations of various analytes within the category, and thereafter present the results of the statistical analysis on those values. To assess the potential ecological significance, exposure concentrations were then compared with the appropriate existing regulatory criteria. A compilation of the various standards, guidelines, and criteria is provided in Table 2. The various regulatory criteria that are summarized in Table 2 are ecologically based and are included in the report to provide a context or benchmark for comparison with environmental exposures to chemicals of emerging concern in the Great Lakes Basin.

## *3.1 Current-use Pesticides*

Pesticides are widely used in both rural and urban applications to control or eliminate organisms that are considered damaging or harmful. Although often used to connote insecticides, the term pesticide actually includes herbicides, fungicides, and various other substances used to control a wide variety of pests. Unlike many of the other chemicals of emerging concern that are addressed in this report, pesticides are among the most stringently regulated substances in the United States and Canada. Before manufacturers can sell a pesticide, they must undergo a thorough scientific evaluation to determine if they can be safely used. Before sanctioning the use of a pesticide, thorough reviews of applications for registration are reviewed by scientific authorities in each government where it is to be used; the most critical evaluations address a variety of potential human health and environmental effects associated with use of the product or products. Potential registrants must generate voluminous scientific data to address many areas of potential concern, including the identity, composition, environmental fate, potential adverse effects and exposure to,

**Table 2** Compilation of various regulatory standards, guidelines, and criteria for the substances examined in this study

| US EPA | | Environment Canada | | | European Union | | | |
|---|---|---|---|---|---|---|---|---|
| Substance | Water (μg/L) | Water (μg/L) | Sediment (mg/kg dwt) | 2° Consumer (mg/kg food) | Water (μg/L) | Sediment (mg/kg dwt) | 2° Consumer (mg/kg food) | References[a] |
| *Current-use pesticides* | | | | | | | | |
| 2,4-D (2,4-dichloro-phenoxyacetic acid) | | 4.0 | | | | | | CCME (2007) |
| Atrazine | | 1.8 | | | | | | CCME (2007) |
| Azinphos-methyl | 0.01 | | | | | | | US EPA (1999) |
| Bromoxynil | | 5.0 | | | | | | CCME (2007) |
| Carbaryl | | 0.2 | | | | | | CCME (2007) |
| Carbofuran | | 1.8 | | | | | | CCME (2007) |
| Chlorpyrifos | 0.041 | 0.02<br>0.002 | | | | | | US EPA (2006);<br>CCME (2008) |
| Diazinon | 0.17 | 0.08 | | | | | | US EPA (2006);<br>IJC (1987) |
| Dicamba | | 10.0 | | | | | | CCME (2007) |
| Diclofop-methyl | | 6.1 | | | | | | CCME (2007) |
| Dimethoate | | 6.2 | | | | | | CCME (2007) |
| Glyphosate | | 65.0 | | | | | | CCME (2007) |
| Malathion | 0.1 | | | | | | | US EPA (2006) |
| MCPA (2-methyl-4-chlorophen-oxyacetic acid) | | 2.6 | | | | | | CCME (2007) |
| Methoprene | | 0.09 | | | | | | CCME (2007) |
| Metolachlor | | 7.8 | | | | | | CCME (2007) |

**Table 2** (continued)

| US EPA | | Environment Canada | | | European Union | | | |
|---|---|---|---|---|---|---|---|---|
| Substance | Water (μg/L) | Water (μg/L) | Sediment (mg/kg dwt) | 2° Consumer (mg/kg food) | Water (μg/L) | Sediment (mg/kg dwt) | 2° Consumer (mg/kg food) | References[a] |
| Metribuzin | | 1.0 | | | | | | CCME (2007) |
| Parathion | 0.013 | | | | | | | US EPA (2006) |
| Picloram | | 29.0 | | | | | | CCME (2007) |
| Simazine | | 10.0 | | | | | | CCME (2007) |
| Triallate | | 0.24 | | | | | | CCME (2007) |
| Trifluralin | | 0.2 | | | | | | CCME (2007) |
| *Organic wastewater contaminants and personal care products* | | | | | | | | |
| 1,4-Dichloro-benzene | | 26.0 | | | 20.0 | 0.94 | | CCME (2007); EUC (2004) |
| Anthracene | | 0.012 | 0.0469 | | 0.1 | 0.203 | | CCME (2007, 2002); EUC (2008e) |
| Benzo(*a*)pyrene | | 0.015 | 0.0319 | | 0.022 | 1.73 | | CCME (2007, 2002); EUC (2008f) |
| Bis (2-ethylhexyl) phthalate | | 16.0 | | | 1.3 | 21.5 | | CCME (2007); EUC (2006a) |
| Bisphenol A | | 0.175 | 0.01 mg/L | | 1.5 | 0.024 mg/kg wwt<br>0.063 mg/kg dwt | 2.67 | EC and HC (2008); EUC (2008a) |
| Fluoranthene | | 0.04 | 0.111 | | 0.01 | 0.071 | | CCME (2007, 2002); EUC (2008f) |
| Naphthalene | | 1.1 | 0.0346 | | 2.0 | 0.373 | | CCME (2007, 2002); EUC (2003d, 2008f) |

**Table 2** (continued)

| US EPA | | Environment Canada | | | European Union | | | |
|---|---|---|---|---|---|---|---|---|
| Substance | Water (µg/L) | Water (µg/L) | Sediment (mg/kg dwt) | 2° Consumer (mg/kg food) | Water (µg/L) | Sediment (mg/kg dwt) | 2° Consumer (mg/kg food) | References[a] |
| Phenanthrene | | 0.4 | 0.0419 | | 1.3 | 2.71 | | CCME (2007, 2002); EUC (2008f) |
| Phenol | | 4.0 | | | 7.7 | 0.233 | | CCME (2007); EUC (2006b) |
| Pyrene | | 0.025 | 0.053 | | 0.0046 | 0.032 | | CCME (2007, 2002); EUC (2008f) |
| Tetrachloroethylene | | 111.0 | | | 51.0 | 0.727 | | CCME (2007); EUC (2005d) |
| *Alkylphenol ethoxylates* | | | | | | | | |
| Nonylphenol (NP) | 6.6 | 1.0 | 1.4 | | 0.33 | 0.039 | 10.0 | CCME (2001); EUC (2002a); US EPA (2005) |
| Nonylphenol ethoxylates (NPEO) | | TEQ[1] | TEQ | | | | | CCME (2001) |
| Nonylphenol ether carboxylates (NPEC) | | TEQ | TEQ | | | | | CCME (2001) |
| Octylphenol (OP) | | TEQ | TEQ | | 0.122 | 0.0074 | 10.0 | CCME (2001); UKEA (2005) |

**Table 2** (continued)

| US EPA | | Environment Canada | | | European Union | | | |
|---|---|---|---|---|---|---|---|---|
| Substance | Water (μg/L) | Water (μg/L) | Sediment (mg/kg dwt) | 2° Consumer (mg/kg food) | Water (μg/L) | Sediment (mg/kg dwt) | 2° Consumer (mg/kg food) | References[a] |
| Octylphenol ethoxylates (OPEO) | | TEQ | TEQ | | | | | CCME (2001) |
| Octylphenol ether carboxylates (OPEC) | | TEQ | TEQ | | | | | CCME (2001) |
| *Synthetic musks* | | | | | | | | |
| Musk ketone | | | | | 6.3 | 0.5 | 0.3 | EUC (2005b) |
| Musk xylene | | | | | 1.1 | 0.3 | 1.0 | EUC (2005a) |
| Acetyl-hexamethyl tetralin (ANTN) | | | | | 2.8 | 1.72 | 1.1 | EUC (2008b) |
| Hexahydro-hexamethylcyclo-pentabenzopyran (HHCB) | | | | | 4.4 | 2.0 | 3.33 | EUC (2008c) |
| *Perfluorinated surfactants* | | | | | | | | |
| Perfluorooctane sulfonic acid (PFOS) | | ENEV[1] 0.491 | | | 2.5 | | 0.0167 | DEFRA (2004); EC (2006a) |
| Perfluorooctanoic acid (PFOA) | | | | | 3.8 | | | Hekster et al. (2002) |

**Table 2** (continued)

| US EPA | | Environment Canada | | | European Union | | | |
|---|---|---|---|---|---|---|---|---|
| Substance | Water (μg/L) | Water (μg/L) | Sediment (mg/kg dwt) | 2° Consumer (mg/kg food) | Water (μg/L) | Sediment (mg/kg dwt) | 2° Consumer (mg/kg food) | References[a] |
| *Brominated flame retardants* | | | | | | | | |
| Penta-bromodiphenyl ether (BDE) | | ENEV 0.053 | ENEV 0.031 | 0.0084 | 0.53 | 0.31 | 1.0 | EC (2006b); EUC (2001) |
| Octa-BDE | | ENEV 0.017 | ENEV 9.1 | 0.06 | > 0.2 | >49 wwt<br>>127 dwt | 6.7 | EC (2006b); EUC (2003b) |
| Deca-BDE | | | ENEV 76.0 | 0.336 | 1.0<br>>0.2 | >148 wwt<br>>384 dwt | 2,500 | EC (2006b) EUC (2002b) |
| Hexabromocyclododecane (HBCD) | | | | | 0.31 | 0.86 mg/kg dwt | 15.3 | EUC (2008g) |
| *Chlorinated paraffins* | | | | | | | | |
| Short-chain chlorinated paraffins (C10–13) (SCCP) | | 0.89 | 3.55 | 10.0 | 0.5 | 0.88 wwt<br>~2.3 dwt | 16.6 | EC (2008); EUC (2008d) |
| Medium-chain chlorinated paraffins (C14–17) (MCCP) | | 1.8 | 27 | 0.42 | 1.0 | 5 wwt<br>~13 dwt | 0.17 | EC (2008); EUC (2005c) |

[a] The following abbreviations are used in the table: *CCME* Canadian Council of Ministers of the Environment, *DEFRA* Department for Environment, Food, and Rural Affairs, *EC* Environment Canada, *EC* and *HC* Environment Canada and Health Canada, *ENEV* Estimated No Effect Value, *EUC* European Commission, *IJC* International Joint Commission, *TEQ* Toxic Equivalent, *UKEA* United Kingdom Environment Agency, *USEPA* United States Environmental Protection Agency

and potential risks associated with using the pesticide. These data allow government authorities to evaluate whether a pesticide has the potential to cause harmful effects in humans and ecosystems, including non-target organisms and endangered species. Government involvement does not end once a pesticide is registered and on the market. Both the registered products and how they are used are monitored through a series of education, compliance, surveillance, and enforcement programs. Pesticides are also periodically re-reviewed to determine if they continue to meet any updated government health and environmental standards, and whether the state of the science indicates they can continue to be used safely.

Concentrations of current-use pesticides in the Great Lakes Basin have been reported in seven studies (Gilliom et al. 2006; Kolpin et al. 2002; Struger and Fletcher 2007; Struger et al. 2004, 2007, 2008; Tertuliana et al. 2008). A number of surveys were conducted by Environment Canada to assess the concentration of pesticides in rural Canadian environments, as well as in the open waters of the Great Lakes. Gilliom et al. (2006) reported the results of an extensive analysis of pesticides in 51 water quality regions of the United States, of which 3 border on the Great Lakes. Kolpin et al. (2002) included the analysis of several pesticides in the recent nationwide reconnaissance of the presence of pharmaceuticals, hormones, and other organic wastewater contaminants in US surface waters. As shown in Fig. 1a and b, the sampling sites included those typical of agricultural and urban drainages, as well as open waters of the Great Lakes. When map coordinates were not provided by the authors, the locations of sampling sites were estimated from figures presented in the papers.

Our analysis focused only on those studies in which concentrations were reported in surface waters and sediments. However, several studies were identified in the literature search, in which analysis of current-use pesticide levels in air and precipitation were described (e.g., Carlson et al. 2004; James and Hites 1999; Miller et al. 2000; Tuduri et al. 2006). In several other studies, atmospheric concentrations and spatial and temporal trends were reported for several historically used organochlorine pesticides (Cortes et al. 1998; Sun et al. 2006a, b).

Struger et al. (2004) summarized the results of exploratory surveys conducted by Environment Canada on the concentrations of current-use pesticides in surface waters of the Laurentian Great Lakes. Large volume (20–50 L) samples were collected, during the period from 1994 to 2000, from Lakes Ontario, Erie, Huron, and Superior. Between 8 and 42 samples were collected each year, with a total of 220 samples collected over the entire investigation. The samples were analyzed for 39 different pesticides, including 15 neutral herbicides, 11 acid herbicides, and 13 organophosphorus insecticides. Because this sampling program was so extensive, the data in the report are presented as ranges (i.e., min–max concentrations). Six analytes, including Barban, diallate-2, triallate, phorate, phosmet, and disulfoton, were not detected in any of the samples. Atrazine, metolachlor, simazine, and 2,4-dichlorophenoxyacetic acid (2,4-D) were detected in greater than 50% of the samples; maximum concentrations reported were 1.04, 0.74, 0.28, and 0.08 μg/L, respectively. Both spatial and seasonal variations in concentrations of the herbicides were observed, which were related to their use patterns. The highest concentrations

A. Surface waters.

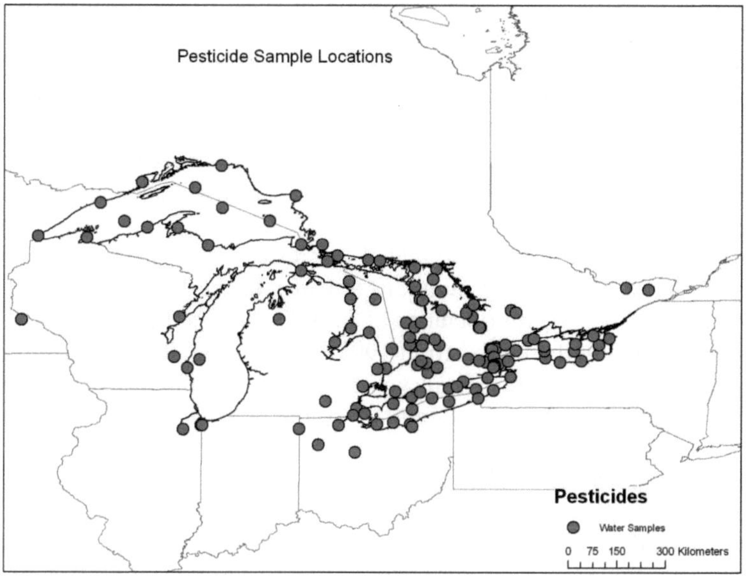


B. Sediments

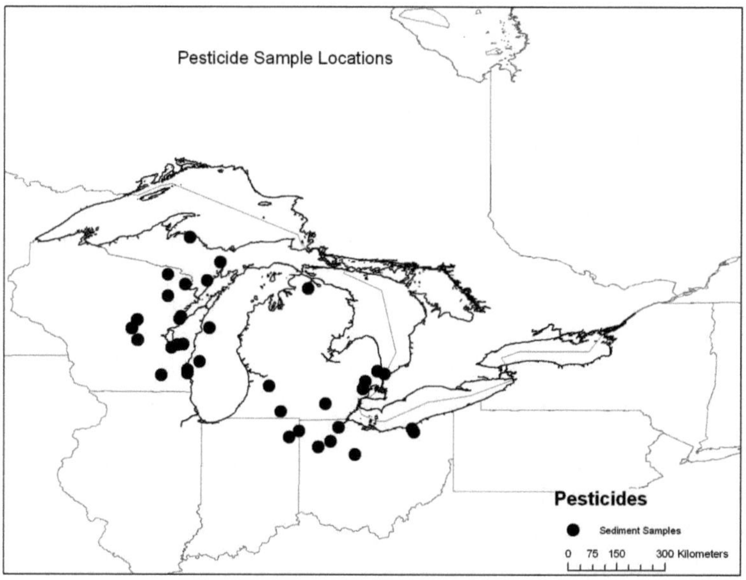


**Fig. 1** Sampling locations for current-use pesticides. **a** Surface waters and **b** sediments

of atrazine, metolachlor, and 2,4-D were detected in samples from the western basin of Lake Erie, because of its close proximity to areas where these pesticides are applied in both agricultural and urban environments. The organophosphorus insecticides were also found primarily in Lake Erie. In general, an increasing concentration gradient from north to south was observed, with residues increasing as follows: Superior < Huron < Ontario < Erie. The authors reported that, although there was evidence of nearfield and farfield influences resulting from use of the various pesticides, all surface water samples met existing water quality guidelines or criteria for the protection of aquatic life.

Struger and Fletcher (2007) presented the results of an extensive monitoring program for lawn care and agricultural pesticides in the Don River and Humber River watersheds, which are tributaries of Lake Ontario in the vicinity of Toronto, ON (Ontario). During the period from 1998 to 2002, a total of 262 samples were collected from the study site. Samples were collected both during base-flow conditions ($n = 139$) and following rainfall events ($n = 123$). The samples were analyzed for a group of 152 pesticide-active ingredients. Because of the extensive amount of data included in the report, only summary data (frequency of detection and maximum concentrations) were available for our analysis. Although 11 different pesticidal ingredients were detected in the two rivers (2,4-D, atrazine, bromacil, carbofuran, chlorpyrifos, cypermethrin, diazinon, dicamba, mecoprop (MCPP), metolachlor, and metribuzin), 72% of the surface water samples contained residues of at least 1 pesticide. Aside from atrazine, which was detected more frequently and at higher concentrations, the study could not statistically distinguish between urban and agricultural pesticide inputs to the watersheds. Concentrations of four pesticides exceeded federal or provincial water quality criteria, in some of the samples. Diazinon exceeded the provincial criteria in 28% of the samples taken. For the other three (atrazine, carbofuran, and chlorpyrifos), the concentrations exceeded the criteria in less than 1% of the samples.

As a result of mosquito control programs implemented because of the increased incidence of the mosquito-borne West Nile Virus, Environment Canada and the Ontario Ministry of the Environment initiated a monitoring program in 2003 to investigate the occurrence and fate of methoprene and its metabolites in source areas and receiving bodies (Struger et al. 2007). Water samples were collected from two tributaries of Hamilton Harbor, four sites in open water areas of Hamilton Harbor, several sites in Cootes Paradise marsh at the western end of Hamilton Harbor, and six sites from a stream near Ottawa, ON. Methoprene was detected in 1 of 37 samples collected in the Hamilton area (0.1 μg/L), and in 1 of the 14 samples collected from the Ottawa stream (0.65 μg/L). Of the 51 samples collected, one exceeded the draft Interim Provincial Water Quality Objective (0.2 μg/L).

Struger et al. (2008) recently reported on the presence of glyphosate in surface waters of southern Ontario. Samples ($n = 502$) were collected from April to December over a 2-year period from 2004 to 2005. The sampling sites selected for the study were typical of agricultural and urban drainages in southern Ontario where glyphosate is commonly used. Because of the extensive information collected, only summary data were presented in the paper. Of the 203 samples collected in 2004 from 26 different field sites, 11 exceeded the detection limit (17 μg/L). When detected, the mean glyphosate concentrations were reported to be in the low μg/L range, with maximum concentrations as high as 40.8 μg/L. Similar results were reported for 2005. Of the 299 samples obtained from 58 different sites, 6 contained glyphosate concentrations above the detection limit. None of the samples collected during the 2-year period exceeded the Canadian Water Quality Guideline (65 μg/L).

In addition to the extensive pesticide monitoring performed throughout Canada and the Great Lakes, Kolpin et al. (2002) analyzed samples from tributaries of Lake Michigan as part of a nationwide reconnaissance program for surface waters in the United States. Samples from seven sites located in Illinois, Michigan, and Wisconsin were collected in 1999–2002 and analyzed for a series of pesticides that included carbaryl, chlordane, chlorpyrifos, diazinon, dieldrin, lindane, and methyl parathion. Of the various analytes, diazinon was detected in a single sample; all other values were reported as being below the detection limit. The concentration of diazinon detected in the sample (0.28 μg/L) exceeded the US Environmental Protection Agency Water Quality Guideline (0.17 μg/L).

Gilliom et al. (2006) presented results of the National Water Quality Assessment Program's first decade of water quality assessments, which included the analysis of 51 major hydrologic systems (study units) across the United States, during the period from 1992 to 2001. Nationally, water samples for pesticide analysis were collected from 186 stream sites. The sampling sites were selected to represent a mixture of agricultural, urban, undeveloped, and mixed-land-use settings. Of the 51 study units included in the analysis, 3 bordered on the Great Lakes. Water samples were analyzed for 75 pesticides and 8 pesticide degradates. In addition, 32 organochlorine pesticide compounds were analyzed in bed sediment samples collected from 1,052 sampling locations across the country. A screening level perspective on the potential significance of pesticide concentrations in streams indicated that 57% of the agricultural streams had concentrations of at least one pesticide that exceeded aquatic life benchmarks at least one time during the year. For urban settings and mixed-land-use settings, 83 and 42% of the streams had concentrations of at least one pesticide that exceeded aquatic life benchmarks at least one time during the year. Pesticides most frequently detected in stream water included five agricultural herbicides (atrazine, metolachlor, cyanazine, alachlor and acetochlor), five urban-use herbicides (simazine, prometon, tebuthiuron, 2,4-D, and diuron), and three widely used insecticides (diazinon, chlorpyrifos, and carbaryl).

Tertuliana et al. (2008) examined the occurrence and distribution of a broad range of organic compounds, including a number of current-use pesticides, in the Tinkers Creek watershed in northeastern Ohio. Sediment samples were collected during May–June of 2006 from 18 locations situated upstream and downstream from wastewater treatment plant discharges to the mainstream and tributaries of Tinkers Creek. Of the six current-use pesticides included in the analysis (atrazine, bromacil, metolachlor, prometon, chlorpyrifos, and diazinon), none were present in the sediment samples above a detection limit of 25 ng/g dry weight (dwt).

Based on the results of 7 studies conducted during the period from 1992 to 2006, a dataset was created containing information on surface water and sediment concentrations for 88 different pesticides. Because of differences in these data, the database was divided and analyzed in two separate sections. Information available from Environment Canada for the concentration of current-use pesticides in Canadian surface waters and the waters of the Great Lakes were provided as summary data (minimum–maximum ranges) in the publications. The database contained 245 summary values reflecting the results for 47 analytes from 4,109 determinations. In contrast, individual data points for single samples were available for all of the United States Geological Survey (USGS; Gilliom et al. 2006) surface water data, representing the results for 74 current-use pesticides and metabolites for 12,226 determinations. The majority of the USGS data covered the sampling period from 1993 to 1997.

Summary statistics for the concentrations of the various analytes in Canadian surface waters and waters from the Great Lakes are presented in Table 3 and are illustrated in Fig. 2. When means and weighted means are given, they refer to the

**Table 3** Summary statistics for current-use pesticides in Canadian waters and the Great Lakes

| Analyte | *n* | Freq det (%) | Mean (μg/L) | Wt Mean (μg/L) | Min (μg/L) | Max (μg/L) |
|---|---|---|---|---|---|---|
| *Neutral herbicides* | | | | | | |
| Atrazine | 358 | 56.3 | 0.7893 | 0.7482 | 0.0047 | 3.6000 |
| Barban | 36 | 0.0 | 0.0038 | 0.0038 | 0.0038 | 0.0038 |
| Benzoylprop-ethyl | 46 | 2.2 | 0.0005 | 0.0004 | 0.0001 | 0.0011 |
| Bromacil | 84 | 14.3 | 0.9650 | 0.6500 | 0.2300 | 1.7000 |
| Butylate | 10 | 20.0 | 0.0019 | 0.0015 | 0.0002 | 0.0035 |
| D-Atrazine | 10 | 100.0 | 0.0098 | 0.0086 | 0.0039 | 0.0157 |
| Deethylatrazine (DEA) | 86 | 1.2 | 0.5125 | 0.7166 | 0.0250 | 1.0000 |
| Diallate total isomer | 36 | 11.1 | 0.0066 | 0.0104 | 0.0033 | 0.0132 |
| Diallate-1 | 10 | 100.0 | 0.0027 | 0.0027 | 0.0025 | 0.0029 |
| Diallate-2 | 10 | 0.0 | 0.0001 | 0.0001 | 0.0001 | 0.0001 |
| Diclofop-methyl | 46 | 17.4 | 0.0008 | 0.0014 | 0.0001 | 0.0020 |
| D-Simazine | 4 | 100.0 | 0.2810 | 0.2810 | 0.2810 | 0.2810 |
| Glyphosate metabolite (AMPA) | 410 | 3.9 | 31.9625 | 32.5890 | 8.5000 | 66.0000 |
| Glyphosate | 415 | 46.3 | 12.2537 | 15.0730 | 1.1700 | 40.8000 |
| Metolachlor | 393 | 50.4 | 0.6702 | 0.7174 | 0.0007 | 1.6000 |
| Metribuzin | 96 | 20.6 | 0.0560 | 0.0861 | 0.0002 | 0.1200 |
| Simazine | 46 | 93.5 | 0.0230 | 0.0302 | 0.0020 | 0.0431 |
| Triallate | 46 | 0.0 | 0.0001 | 0.0000 | 0.0000 | 0.0004 |
| Trifluralin | 46 | 10.9 | 0.0003 | 0.0006 | 0.0000 | 0.0009 |
| *Acid herbicides* | | | | | | |
| 2,3,6-Trichloro-benzoic acid (2,3,6-TBA) | 58 | 34.5 | 0.0168 | 0.0158 | 0.0002 | 0.0511 |

**Table 3** (continued)

| Analyte | *n* | Freq det (%) | Mean (μg/L) | Wt Mean (μg/L) | Min (μg/L) | Max (μg/L) |
|---|---|---|---|---|---|---|
| 2,4,5-Trichloro-phenoxyacetic acid (2,4,5-T) | 58 | 20.7 | 0.0012 | 0.0010 | 0.0002 | 0.0024 |
| 2,4,5-Trichloro-phenoxyproprionic acid (2,4,5-TP) | 58 | 27.6 | 0.0044 | 0.0049 | 0.0002 | 0.0108 |
| 2,4-D | 228 | 32.6 | 0.6629 | 1.0424 | 0.0014 | 1.6000 |
| 2,4-Dichloro-phenoxybutanoic acid (2,4-DB) | 58 | 25.9 | 0.0072 | 0.0133 | 0.0009 | 0.0243 |
| 2,4-Dichloro-phenoxyproprionic acid (2,4-DP) | 58 | 6.9 | 0.0029 | 0.0044 | 0.0002 | 0.0064 |
| Bromoxynil | 58 | 22.4 | 0.0015 | 0.0014 | 0.0002 | 0.0027 |
| Dicamba | 228 | 21.6 | 0.4511 | 0.5954 | 0.0002 | 2.2000 |
| MCPA | 58 | 17.2 | 0.0026 | 0.0029 | 0.0002 | 0.0054 |
| Methylchloro-phenoxybutanoic acid (MCPB) | 58 | 1.7 | 0.0003 | 0.0003 | 0.0002 | 0.0007 |
| Methylchloro-phenoxypropionic acid (MCPP) | 170 | 50.6 | 1.6500 | 1.8165 | 1.1000 | 2.4000 |
| Picloram | 58 | 12.1 | 0.0023 | 0.0038 | 0.0003 | 0.0061 |
| *Insecticides* | | | | | | |
| Azinphos-methyl | 24 | 16.7 | 0.0155 | 0.0210 | 0.0006 | 0.0452 |
| Carbofuran | 86 | 1.2 | 0.5250 | 0.7238 | 0.0500 | 1.0000 |
| Chlorpyrifos | 108 | 6.8 | 0.1370 | 0.1299 | 0.0002 | 0.5200 |
| Cypermethrin | 84 | 1.2 | 0.2025 | 0.2786 | 0.0250 | 0.3800 |
| Diazinon | 194 | 39.6 | 0.3038 | 0.6169 | 0.0003 | 1.0000 |
| Dibrom | 24 | 8.3 | 0.0035 | 0.0032 | 0.0003 | 0.0078 |
| Dimethoate | 24 | 8.3 | 0.0117 | 0.0159 | 0.0003 | 0.0344 |
| Disulfoton | 24 | 0.0 | 0.0002 | 0.0002 | 0.0002 | 0.0002 |
| Ethion | 24 | 4.2 | 0.0007 | 0.0010 | 0.0002 | 0.0019 |
| Fonofos | 24 | 4.2 | 0.0010 | 0.0014 | 0.0001 | 0.0028 |
| Malathion | 24 | 8.3 | 0.0070 | 0.0096 | 0.0001 | 0.0208 |
| Methoprene | 37 | 2.7 | 0.0250 | 0.0162 | 0.0000 | 0.1000 |
| Parathion | 24 | 8.3 | 0.0014 | 0.0019 | 0.0003 | 0.0038 |
| Phorate | 24 | 0.0 | 0.0001 | 0.0001 | 0.0001 | 0.0001 |
| Phosmet | 24 | 0.0 | 0.0006 | 0.0006 | 0.0006 | 0.0006 |
| Terbufos | 24 | 8.3 | 0.0048 | 0.0066 | 0.0001 | 0.0143 |

analysis of maximum concentration values reported by the authors. The neutral herbicides atrazine, diallate, metolachlor, and simazine were detected in 50–100% of the samples. Other herbicides frequently detected were glyphosate (46.3%) and several of the phenoxy acid or benzoic acid herbicides, including 2,3,6-trichlorobenzoic acid (2,3,6-TBA; 34.5%), 2,4-D (32.6%), dicamba (21.6%), and

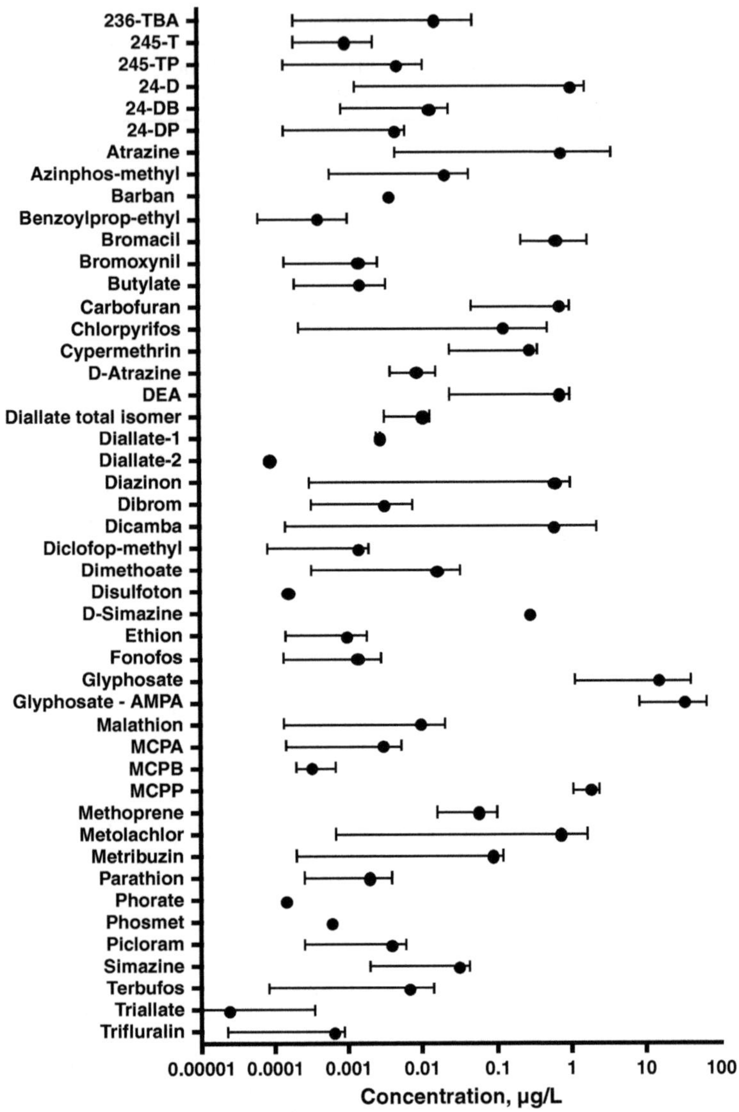


**Fig. 2** Minimum, weighted mean, and maximum concentrations reported for current-use pesticides in surface waters, Canadian streams, and Great Lakes

MCPP (mecoprop; 50.6%). For these herbicides, the maximum concentrations were generally in the low microgram per liter range (0.01–3.6 μg/L), with the exception of glyphosate, in which a maximum concentration of 40.8 μg/L was reported for a single sample. The remaining herbicides were all detected at lower frequencies and at concentrations less than 1 μg/L. When compared with current water quality guidelines or criteria (Table 2), atrazine exceeded the standards in some samples (<1%).

Of the insecticides detected in Canadian surface waters and waters of the Great Lakes, azinphos-methyl and diazinon were among those most frequently detected (16.7 and 39.6%) with maximum concentrations ranging from 0.045 to 1.0 μg/L (Table 3). The remaining insecticides were detected less frequently (<10%) and at maximum concentrations less than 1.0 μg/L. Disulfoton, phorate, and phosmet were not detected in any of the samples. As summarized previously (Struger and Fletcher 2007), diazinon exceeded current regulatory criteria in 28% of the samples taken, whereas azinphos-methyl, carbofuran, chlorpyrifos, and methoprene seldom (<1%) exceeded the standard, and when they did so, it was usually in a single sample.

Summary statistics for the concentrations of the various pesticides in surface waters from three US water quality units that border the Great Lakes are presented in Table 4. Of the neutral herbicides, the highest maximum concentrations (4.7–85.20 μg/L) were reported for acetochlor, alachlor, atrazine, bentazon, cyanazine, fluometuron, metolachlor, metribuzin, propham, and simazine. Maximum concentrations ranging from 3.05 to 37.3 μg/L were also observed for several phenoxy acid herbicides. In contrast, the 95th percentile concentrations for all of the herbicides were less than 1.0 μg/L, with the exception of atrazine (2.77 μg/L) and 2,4-D (1.178 μg/L). Maximum concentrations, ranging from 2.57 to 22.5 μg/L, were reported for several insecticides (chlorpyrifos, diazinon, malathion, methiocarb, and oxamyl), although the 95th percentile concentrations for these insecticides were less than 1.0 μg/L. For the remaining fungicides and parasiticides, all of the reported concentrations were below 1.0 μg/L.

When the concentrations of current-use pesticides in US surface waters were compared to current water quality guidelines or criteria (Table 2), the reported concentrations for bromoxynil, carbaryl, carbofuran, dicamba, picloram, simazine, triallate, and trifluralin were all within the regulatory criteria. With 2,4-D, malathion, 2-methyl-4-chlorophenoxyacetic acid (MCPA), metolachlor, and metribuzin, the 95th percentile concentrations were below the regulatory guidelines, but exceedances were noted for the maximum reported concentrations. Concentrations of atrazine, azinphos-methyl, chlorpyrifos, diazinon, and parathion exceeded the regulatory standards in 6–32% of the samples.

As summarized in Table 5, limited data were available for the concentrations of current-use pesticides in sediments; the concentrations reported by Gilliom et al. (2006) and Tertuliana et al. (2008) are presented in this same table. Of the 11 current-use pesticides that have been analyzed, 5 were detected in 18–27% of the samples ($n = 38$) at concentrations ranging from 0.5 to 13.0 ng/g dwt.

## 3.2 Pharmaceuticals

Prescription and non-prescription drugs are extensively used to treat diseases in humans and domestic animals. In urban environments, these compounds make their way to wastewater treatment plants where they may be ultimately discharged to the environment in effluents or through application onto agricultural land as biosolids.

**Table 4** Summary statistics for current-use pesticides in US streams

| Analyte | *n* | Freq det (%) | 50th percentile (μg/L) | 95th percentile (μg/L) | Min (μg/L) | Max (μg/L) |
|---|---|---|---|---|---|---|
| *Neutral herbicides* | | | | | | |
| Acetochlor | 165 | 100 | 0.030 | 0.142 | 0.001 | 10.60 |
| Alachlor | 165 | 100 | 0.008 | 0.176 | 0.001 | 6.70 |
| Atrazine | 165 | 100 | 0.041 | 2.770 | 0.001 | 85.20 |
| Benfluralin | 165 | 76 | 0.005 | 0.011 | 0.001 | 0.016 |
| Bentazon | 165 | 100 | 0.030 | 0.130 | 0.003 | 7.32 |
| Bromacil | 164 | 100 | 0.040 | 0.120 | 0.001 | 1.63 |
| Butylate | 165 | 100 | 0.001 | 0.040 | 0.001 | 0.100 |
| Cyanazine | 165 | 100 | 0.011 | 0.777 | 0.001 | 9.97 |
| Deethylatrazine (DEA) | 164 | 99 | 0.022 | 0.393 | 0.001 | 3.87 |
| Dichlobenil | 165 | 95 | 0.040 | 0.066 | 0.003 | 0.210 |
| Dichlorprop | 165 | 96 | 0.050 | 0.070 | 0.003 | 0.600 |
| Dinoseb | 165 | 97 | 0.040 | 0.060 | 0.001 | 0.210 |
| Diuron | 165 | 98 | 0.040 | 0.060 | 0.001 | 0.180 |
| Ethyldipropyl-thiocarbamate (EPTC) | 165 | 76 | 0.002 | 0.020 | 0.001 | 0.020 |
| Ethalfluralin | 165 | 76 | 0.005 | 0.017 | 0.001 | 0.018 |
| Fenuron | 165 | 99 | 0.030 | 0.130 | 0.001 | 0.210 |
| Fluometuron | 165 | 100 | 0.030 | 0.085 | 0.002 | 5.00 |
| Linuron | 165 | 76 | 0.003 | 0.018 | 0.001 | 0.394 |
| Metolachlor | 165 | 100 | 0.012 | 0.774 | 0.001 | 77.60 |
| Metribuzin | 165 | 76 | 0.003 | 0.163 | 0.002 | 4.92 |
| Molinate | 165 | 76 | 0.002 | 0.009 | 0.001 | 0.010 |
| Napropamide | 164 | 77 | 0.007 | 0.020 | 0.001 | 0.056 |
| Neburon | 165 | 99 | 0.030 | 0.240 | 0.001 | 0.240 |
| Norflurazon | 165 | 99 | 0.021 | 0.130 | 0.001 | 0.240 |
| Pebulate | 165 | 76 | 0.005 | 0.008 | 0.001 | 0.091 |
| Pendimethalin | 164 | 77 | 0.003 | 0.011 | 0.001 | 0.113 |
| Prometon | 165 | 99 | 0.013 | 0.093 | 0.001 | 0.681 |
| Pronamide | 165 | 76 | 0.009 | 0.011 | 0.001 | 0.021 |
| Propachlor | 166 | 99 | 0.005 | 0.034 | 0.001 | 0.170 |
| Propanil | 165 | 76 | 0.005 | 0.021 | 0.001 | 0.080 |
| Propham | 165 | 100 | 0.110 | 0.110 | 0.001 | 5.61 |
| Simazine | 161 | 94 | 0.018 | 0.996 | 0.001 | 4.70 |
| Tebuthiuron | 165 | 76 | 0.002 | 0.011 | 0.001 | 0.024 |
| Terbacil | 165 | 76 | 0.017 | 0.018 | 0.001 | 0.106 |
| Thiobencarb | 164 | 76 | 0.002 | 0.011 | 0.001 | 0.011 |
| Triallate | 165 | 76 | 0.002 | 0.005 | 0.001 | 0.021 |
| Trifluralin | 165 | 76 | 0.005 | 0.050 | 0.001 | 0.050 |
| *Acid herbicides* | | | | | | |
| 2,4,5-T | 165 | 100 | 0.040 | 0.040 | 0.001 | 0.210 |
| 2,4,5-TP | 165 | 100 | 0.030 | 0.110 | 0.001 | 3.050 |
| 2,4-D | 165 | 100 | 0.080 | 1.178 | 0.001 | 6.140 |
| 2,4-DB | 165 | 100 | 0.130 | 0.260 | 0.003 | 28.60 |

**Table 4** (continued)

| Analyte | *n* | Freq det (%) | 50th percentile (μg/L) | 95th percentile (μg/L) | Min (μg/L) | Max (μg/L) |
|---|---|---|---|---|---|---|
| Acifluorfen | 165 | 76 | 0.040 | 0.240 | 0.003 | 0.240 |
| Bromoxynil | 165 | 76 | 0.100 | 0.275 | 0.016 | 0.275 |
| Chloramben methyl | 165 | 76 | 0.005 | 0.110 | 0.002 | 3.50 |
| Clopyralid | 165 | 92 | 0.070 | 0.240 | 0.003 | 0.240 |
| Dacthal | 164 | 76 | 0.002 | 0.020 | 0.001 | 0.144 |
| Dicamba | 165 | 100 | 0.050 | 0.100 | 0.001 | 0.635 |
| MCPA | 165 | 100 | 0.100 | 0.130 | 0.002 | 37.30 |
| MCPB | 165 | 100 | 0.130 | 0.510 | 0.002 | 19.00 |
| Picloram | 165 | 100 | 0.040 | 0.140 | 0.001 | 0.240 |
| Triclopyr | 165 | 100 | 0.040 | 0.140 | 0.001 | 0.850 |
| *Insecticides* | | | | | | |
| Aldicarb | 165 | 76 | 0.100 | 0.600 | 0.040 | 0.600 |
| Azinphos-methyl | 164 | 73 | 0.003 | 0.020 | 0.001 | 0.124 |
| Carbaryl | 172 | 73 | 0.002 | 0.021 | 0.001 | 0.030 |
| Carbofuran | 165 | 76 | 0.009 | 0.010 | 0.001 | 0.011 |
| Chlorpyrifos | 172 | 96 | 0.003 | 0.339 | 0.002 | 10.00 |
| Diazinon | 172 | 97 | 0.006 | 0.186 | 0.001 | 22.50 |
| Disulfoton | 165 | 76 | 0.001 | 0.011 | 0.001 | 0.021 |
| Ethoprop | 165 | 76 | 0.002 | 0.009 | 0.001 | 0.010 |
| Fonofos | 165 | 100 | 0.001 | 0.050 | 0.001 | 0.843 |
| Malathion | 165 | 100 | 0.014 | 0.078 | 0.001 | 6.40 |
| Methiocarb | 165 | 100 | 0.030 | 0.130 | 0.003 | 2.57 |
| Methomyl | 165 | 99 | 0.030 | 0.240 | 0.001 | 0.240 |
| Methyl parathion | 172 | 73 | 0.002 | 0.030 | 0.001 | 0.330 |
| Oxamyl | 165 | 100 | 0.080 | 0.080 | 0.002 | 3.70 |
| Parathion | 165 | 100 | 0.005 | 0.056 | 0.001 | 0.080 |
| *cis*-Permethrin | 164 | 43 | 0.004 | 0.040 | 0.001 | 0.177 |
| Phorate | 165 | 76 | 0.006 | 0.018 | 0.001 | 0.510 |
| Propargite | 164 | 77 | 0.011 | 0.020 | 0.001 | 0.056 |
| Propoxur | 165 | 100 | 0.060 | 0.060 | 0.003 | 0.595 |
| Terbufos | 165 | 76 | 0.002 | 0.020 | 0.001 | 0.020 |
| *Fungicides* | | | | | | |
| Chlorothalonil | 165 | 81 | 0.070 | 0.100 | 0.003 | 0.240 |
| Dinitro-*ortho*-cresol (DNOC) | 165 | 98 | 0.060 | 0.130 | 0.002 | 0.190 |
| *Parasiticide* | | | | | | |
| Oryzalin | 165 | 100 | 0.021 | 0.140 | 0.001 | 0.140 |

**Table 5** Summary statistics for current-use pesticides in US sediments

| Analyte | *n* | Freq det (%) | Mean (ng/g dwt) | Min (ng/g dwt) | Max (ng/g dwt) |
|---|---|---|---|---|---|
| *Neutral herbicides* | | | | | |
| Atrazine | 18 | 0 | 12.5 | 12.5 | 12.5 |
| Bromacil | 18 | 0 | 12.5 | 12.5 | 12.5 |
| Metolachlor | 18 | 0 | 12.5 | 12.5 | 12.5 |
| Prometon | 18 | 0 | 12.5 | 12.5 | 12.5 |
| *Acid herbicides* | | | | | |
| Dacthal | 38 | 18 | 4.2 | 2.5 | 13.0 |
| *Insecticides* | | | | | |
| Chlorpyrifos | 18 | 0 | 12.5 | 12.5 | 12.5 |
| Diazinon | 18 | 0 | 12.5 | 12.5 | 12.5 |
| Endosulfan | 37 | 22 | 1.0 | 0.5 | 5.0 |
| *cis*-Permethrin | 38 | 18 | 0.8 | 0.5 | 2.5 |
| *trans*-Permethrin | 37 | 27 | 4.7 | 2.5 | 13.0 |
| *Fungicides* | | | | | |
| Chloroneb | 38 | 18 | 4.2 | 2.5 | 13.0 |

In rural environments, the drugs that are used by humans or administered to livestock and poultry have the potential to enter the aquatic environment through runoff into surface water or through contamination of groundwater.

The detection of pharmaceutical compounds in the aquatic environment has received considerable attention in the technical literature and in the popular press. Prescription and non-prescription drugs have been detected in rivers and streams in both North America and Europe, and several recent reviews have addressed the occurrence and fate of pharmaceuticals in the environment (e.g., Monteiro and Boxall 2010; Walraven and Laane 2008). Wastewater treatment plants and agricultural operations are important sources of these contaminants to surface waters, but limited data exist on the spatial distribution of drugs in receiving waters near such sources. Concentrations of pharmaceuticals in surface waters from the Great Lakes Basin have been reported in 11 studies. The locations of the sampling sites for pharmaceuticals are shown in Fig. 3.

Kolpin et al. (2002) reported the results of a nationwide reconnaissance study of the occurrence of pharmaceuticals, hormones, and other organic wastewater contaminants in water resources. Samples were collected during 1999 and 2000 from a network of 139 streams across 30 states, of which 19 sites were located within the Lake Michigan drainage basin. The selection of the sampling sites was biased toward areas considered to be susceptible to contamination from human, industrial, and agricultural sources (i.e., downstream from urbanization, livestock production, etc.). Of the 95 organic wastewater contaminants included in the study, 41 represented different pharmaceuticals, including veterinary care products and human prescription and non-prescription drugs. In general, the human and veterinary antibiotics were detected less frequently (14 of 31 were non-detected in all samples) than

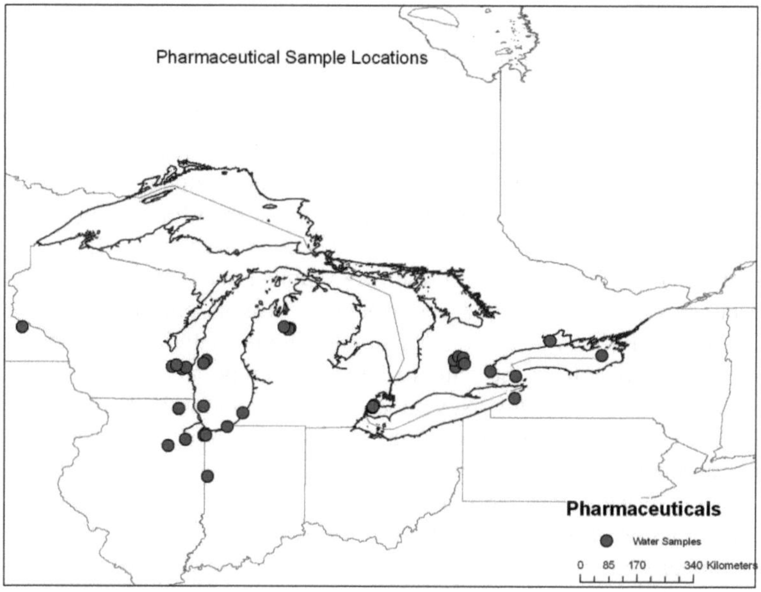


**Fig. 3** Sampling locations for pharmaceutical compounds

were the other prescription drugs (5 of 15 were non-detected). Detection limits for the various analytes ranged from 0.001 to 0.26 μg/L (mean ± SD; 0.065 ± 0.054). Non-prescription drugs (including caffeine and cotinine) were generally detected at higher frequencies, than were other pharmaceuticals, which may be at least partially a result of their greater annual use.

Metcalfe et al. (2003) analyzed for the presence of 18 pharmaceuticals in surface waters from the lower Great Lakes. Samples were collected in the summer and fall of 2000 at open water sites in Lake Erie and Lake Ontario. Additional samples were collected in 2000 and 2002 from sites in the Detroit River and Hamilton Harbor in the vicinity of discharges from wastewater treatment plants for the cities of Windsor and Hamilton, Ontario. A variety of drugs were detected at nanogram per liter concentrations from sites that were up to 500 m downstream from the discharges of the wastewater treatment plants. The highest concentrations were noted for ibuprofen (0.79 μg/L), naproxen (0.55 μg/L) and carbamazepine (0.65 μg/L). Caffeine and cotinine (nicotine metabolite) were generally present in the receiving waters and are considered markers for human excretion. Pharmaceutical compounds were not detected (<1 to <10 ng/L) at open water locations in western Lake Erie or in the Niagara River near the municipality of Niagara-on-the-Lake. However, clofibric acid, ketoprofen, fenoprofen, and carbamazepine were detected at low levels (0.02–0.06 μg/L) in samples collected in 2000 from Lake Ontario and at sites on the Niagara River near Fort Erie that were relatively remote from wastewater discharges.

Jones-Lepp (2006) addressed the utility of analyzing for chemical markers of human waste contamination in surface waters and the development of analytical methods for detecting urobilin, a by-product of hemoglobin degradation. Surface water samples were collected in 2004 from various sites in the United States, including two samples from recreational beaches on the Lake Michigan shoreline that were near wastewater treatment plant effluent outfalls. The samples were analyzed for urobilin and azithromycin, an antibiotic widely prescribed for human use. Neither analyte was present in the samples above the detection limit (<1 ng/L).

Lissemore et al. (2006) described the geographical and temporal distribution of pharmaceuticals detected within seven tributaries of the Grand River watershed in southern Ontario. Related work, which summarized the analytical methods and their application to samples from the same locations, was reported by Hao et al. (2006). Samples were collected in May–November of 2003 and March–April of 2004 from seven agricultural and one urban location. The rural sampling sites were previously identified as having had a high risk for agricultural runoff. Of 28 pharmaceuticals surveyed, 14 were detected in the stream samples ($n = 125$) at nanogram per liter concentrations. The five most frequently detected analytes were lincomycin (91.2%), monensin (75.2%), carbamazepine (48.8%), sulfamethazine (32.8%), and trimethoprim (20%); median and maximum concentrations for these analytes ranged from 1 to 44 ng/L and from 15 to 1,172 ng/L, respectively. The highest concentrations were reported for monensin and sulfamethazine; these are used strictly for livestock production, which is consistent with the selection of sampling sites in areas suspected of agricultural contamination. With the exception of carbamazepine (0.16–24 ng/L), prescription drugs with strictly human uses were not detected in the samples.

Servos et al. (2007) examined the presence of eight pharmaceuticals and the antimicrobial, triclosan, in drinking water sources used in Southern Ontario. Samples were collected during the fall of 2002 from facilities located near Burlington, ON. Of the facilities included in the study, 4 used river waters and 11 used lake waters as source waters. Most of the river water sampling sites were situated downstream from one or more municipal wastewater treatment plants, whereas the lake sites were more remote from any direct discharges. River water samples showed the highest levels of naproxen and ibuprofen, with concentrations ranging as high as 176 and 150 ng/L, respectively. Low levels of gemfibrozil (19.2 ng/L), diclofenac (15 ng/L), and indomethacin (6 ng/L) were also detected in river water. In contrast, the concentrations of the various pharmaceuticals in lake water samples were less than 5 ng/L.

Chu and Metcalfe (2007) described the development of sensitive analytical methods and their application to the analysis of several selective serotonin re-uptake inhibitors that are widely used to treat depression. Fish ($n = 7$) were collected in 2002 from Hamilton Harbor. Paroxetine, fluoxetine, and norfluoxetine were detected in 54–86% of the fish with maximum concentrations ranging from 0.58 to 1.08 ng/g wet weight (wwt).

Tertuliana et al. (2008) examined the occurrence and distribution of 151 pharmaceutical and organic wastewater compounds in the Tinkers Creek watershed in northeastern Ohio. Tinkers Creek is the largest tributary leading into the Cuyahoga River. The study included an evaluation of passive samplers, such as polar organic chemical integrative samplers (POCIS) and semi-permeable membrane devices (SPMD) for providing time-weighted measures of concentration in the streams over the exposure period. Additional sediment samples were also collected at each of the stations where the passive samplers were deployed. The samples were collected during May–June of 2006 from 18 locations situated upstream and downstream from wastewater treatment plant discharges to the mainstream and tributaries of Tinkers Creek. Because of variations in site-specific factors (temperature and water velocity) and the strong sorptive capacity, the concentrations of the various analytes were reported in units of nanograms per POCIS disk and could not be adjusted for variations in sampling rates. For this reason, results of the water analysis provided qualitative indications of the presence/absence of the various analytes at the sampling locations. Of the 52 pharmaceuticals included in the analysis, a total of 12 antibiotics and 20 prescription and non-prescription drugs were detected in water at one or more of the sites. The detections were generally higher in passive samplers located downstream of outfalls as compared to those deployed upstream of outfalls or at reference locations. Additionally, eight pharmaceuticals were detected in streambed sediments at all sites in the Tinkers Creek watershed. Median and maximum concentrations for the various pharmaceutical compounds in sediment ranged from 2 to 13 ng/g dwt and from 3.3 to 75 ng/g dwt, respectively. Non-prescription drugs including diphenhydramine, caffeine, and miconazole were detected at higher frequencies (28–50%) in the sediments at concentrations ranging from 0.3 to 75 ng/g dwt.

Three additional studies have reported concentrations for a few pharmaceuticals in a limited number of samples taken from the Detroit River. The sampling sites ranged from locations downstream of the Little River Wastewater Treatment Plant (Windsor, ON) to further downstream near the intake site for the A.H. Weeks Water Treatment Plant, which furnishes drinking water to the city of Windsor, ON. Boyd et al. (2003) reported concentrations for clofibric acid (103 ng/L), ibuprofen (<2.6 ng/L), fluoxetine (<25.4 ng/L), and naproxen (63 ng/L) in samples of raw intake water taken in 2002. Hua et al. (2006a) reported trace levels of carbamazepine (0.3–3.8 ng/L), caffeine (2.3–24 ng/L), and cotinine (0.1–1.6 ng/L) in raw intake water samples collected in 2003. The primary focus of these investigations was on the effectiveness of water treatment technologies for the removal of trace contaminants. In a follow-up study, Hua et al. (2006b) reported on the spatial distribution and the influence of seasonal changes on the concentration of a variety of acidic and neutral pharmaceuticals in samples from the Detroit River. Ten of 17 compounds were detected downstream of the wastewater treatment plant at the confluence of the Little River and the Detroit River at concentrations of less than 100 ng/L, on average. In spite of the large dilution, three compounds (carbamazepine, caffeine, and cotinine) were detectable at low nanogram per liter concentrations at shoreline sites several kilometers downstream.

Based on the available information, a database was created containing 1,021 data points (either single values, means, or ranges) for 60 pharmaceutical compounds analyzed in surface water, sediments, and biota from the Great Lakes Basin. As summarized in Table 6, the number of surface water samples analyzed for each of the substances varies from 2 to 303. The frequency with which the various compounds have been detected varies from 0 to as high as 84%. Detection limits among the compounds varied from 0.001 to 0.26 μg/L. In general, higher detection limits were reported by Kolpin et al. (2002), which may derive from the analytical methods employed being designed to detect multiple analytes at the sacrifice of some sensitivity.

**Table 6** Summary statistics for pharmaceutical compounds in surface waters

| Analyte | *n* | Freq det (%) | Mean (μg/L) | Wt Mean (μg/L) | Min (μg/L) | Max (μg/L) |
|---|---|---|---|---|---|---|
| *Veterinary and human antibiotics* | | | | | | |
| Azithromycin | 16 | 0 | 0.0005 | 0.0005 | 0.0005 | 0.0005 |
| Carbadox | 143 | 0 | 0.0260 | 0.0196 | 0.0100 | 0.0500 |
| Chloramphenicol | 125 | 0 | 0.0025 | 0.0025 | 0.0025 | 0.0025 |
| Chlortetracycline | 159 | 1 | 0.1247 | 0.1632 | 0.0250 | 0.1920 |
| Ciprofloxacin | 18 | 0 | 0.0158 | 0.0158 | 0.0100 | 0.0250 |
| Doxycycline | 143 | 3 | 0.0556 | 0.0689 | 0.0250 | 0.0730 |
| Enrofloxacin | 18 | 0 | 0.0256 | 0.0256 | 0.0100 | 0.0500 |
| Erythromycin | 231 | 29 | 0.0504 | 0.0396 | 0.0001 | 0.2800 |
| Lincomycin | 231 | 84 | 0.0613 | 0.2060 | 0.0140 | 0.3550 |
| Monensin | 213 | 80 | 0.2328 | 0.7355 | 0.0200 | 1.1720 |
| Norfloxacin | 18 | 0 | 0.0158 | 0.0158 | 0.0100 | 0.0250 |
| Oxytetracycline | 159 | 0 | 0.0332 | 0.0159 | 0.0030 | 0.1000 |
| Roxithromycin | 143 | 6 | 0.0103 | 0.0041 | 0.0020 | 0.0250 |
| Sarafloxacin | 18 | 0 | 0.0100 | 0.0100 | 0.0100 | 0.0100 |
| Sulfachloropyridazine | 229 | 15 | 0.0542 | 0.0301 | 0.0070 | 0.0700 |
| Sulfadimethoxine | 159 | 8 | 0.0405 | 0.0494 | 0.0250 | 0.0560 |
| Sulfamerazine | 159 | 0 | 0.0126 | 0.0055 | 0.0002 | 0.0250 |
| Sulfamethazine | 247 | 43 | 0.0569 | 0.2169 | 0.0005 | 0.4100 |
| Sulfamethizole | 18 | 0 | 0.0500 | 0.0500 | 0.0500 | 0.0500 |
| Sulfamethoxazole | 150 | 6 | 0.0260 | 0.0147 | 0.0090 | 0.0990 |
| Sulfathiazole | 159 | 3 | 0.0271 | 0.0208 | 0.0160 | 0.0500 |
| Tetracycline | 159 | 0 | 0.0437 | 0.0358 | 0.0250 | 0.1000 |
| Trimethoprim | 253 | 41 | 0.0161 | 0.0128 | 0.0002 | 0.1340 |
| Tylosin | 143 | 3 | 0.0150 | 0.0075 | 0.0050 | 0.0300 |
| Virginiamycin | 143 | 0 | 0.0325 | 0.0194 | 0.0150 | 0.0500 |
| *Prescription drugs* | | | | | | |
| Albuterol | 7 | 0 | 0.0145 | 0.0145 | 0.0145 | 0.0145 |
| Atorvastatin | 12 | 50 | 0.0088 | 0.0088 | 0.0050 | 0.0150 |
| Bezafibrate | 189 | 13 | 0.0423 | 0.0210 | 0.0010 | 0.2000 |
| Carbamazepine | 290 | 70 | 0.0724 | 0.0596 | 0.00003 | 0.6600 |
| Cimetidine | 7 | 29 | 0.0118 | 0.0118 | 0.0035 | 0.0530 |
| Clofibric acid | 204 | 19 | 0.0285 | 0.0213 | 0.0005 | 0.1750 |
| Codeine | 9 | 0 | 0.0527 | 0.0527 | 0.0120 | 0.1000 |
| Cyclophosphamide | 15 | 20 | 0.0029 | 0.0040 | 0.0002 | 0.0050 |
| Dehydronifedipine | 7 | 14 | 0.0059 | 0.0059 | 0.0050 | 0.0110 |
| Diclofenac | 208 | 17 | 0.0331 | 0.0253 | 0.0002 | 0.1940 |

**Table 6** (continued)

| Analyte | *n* | Freq det (%) | Mean (μg/L) | Wt Mean (μg/L) | Min (μg/L) | Max (μg/L) |
|---|---|---|---|---|---|---|
| Digoxin | 2 | 0 | 0.1300 | 0.1300 | 0.1300 | 0.1300 |
| Digoxigenin | 7 | 0 | 0.0040 | 0.0040 | 0.0040 | 0.0040 |
| Diltiazem | 7 | 14 | 0.0084 | 0.0084 | 0.0060 | 0.0230 |
| Enalaprilat | 7 | 0 | 0.0760 | 0.0760 | 0.0760 | 0.0760 |
| Fenoprofen | 78 | 36 | 0.0335 | 0.0359 | 0.0002 | 0.1420 |
| Fluoxetine | 23 | 26 | 0.0142 | 0.0139 | 0.0050 | 0.0460 |
| Gemfibrozil | 303 | 26 | 0.0157 | 0.0122 | 0.0010 | 0.1120 |
| Indomethacin | 159 | 16 | 0.0047 | 0.0048 | 0.0002 | 0.0180 |
| Meclocycline sulfosalicylate | 125 | 0 | 0.0025 | 0.0025 | 0.0025 | 0.0025 |
| Metformin | 7 | 14 | 0.0213 | 0.0213 | 0.0015 | 0.1400 |
| Norfluoxetine | 15 | 0 | 0.0048 | 0.0049 | 0.0045 | 0.0050 |
| Paroxetine metabolite | 7 | 0 | 0.1300 | 0.1300 | 0.1300 | 0.1300 |
| Pentoxifylline | 15 | 40 | 0.0039 | 0.0054 | 0.0001 | 0.0090 |
| Ranitidine | 7 | 0 | 0.0050 | 0.0050 | 0.0050 | 0.0050 |
| Warfarin | 7 | 0 | 0.0005 | 0.0005 | 0.0005 | 0.0005 |
| *Non-prescription drugs* | | | | | | |
| Acetaminophen | 7 | 0 | 0.0045 | 0.0045 | 0.0045 | 0.0045 |
| Caffeine | 46 | 68 | 0.0316 | 0.0448 | 0.0005 | 0.3000 |
| Cotinine | 43 | 72 | 0.0074 | 0.0106 | 0.0001 | 0.0500 |
| 1,7-Dimethylxanthine | 7 | 0 | 0.0090 | 0.0090 | 0.0090 | 0.0090 |
| Ibuprofen | 211 | 16 | 0.0729 | 0.0688 | 0.0013 | 0.7900 |
| Ketoprofen | 83 | 33 | 0.0140 | 0.0227 | 0.0001 | 0.0500 |
| Naproxen | 297 | 26 | 0.0630 | 0.0712 | 0.0025 | 0.5510 |

Of the 57 different compounds analyzed in the water samples, 24 were not detected in any of the samples. With respect to the 980 data points for water, detectable concentrations of the various pharmaceutical compounds were reported for 330 samples (34%). Veterinary and human antibiotics were generally the least-often detected pharmaceutical compounds. Of those that were detected, the most frequent include erythromycin (29%), lincomycin (84%), monensin (80%), sulfamethazine (43%), and trimethoprim (41%); mean and maximum reported concentrations for these analytes ranged from 0.02 to 0.24 μg/L and from 0.13 to 1.17 μg/L, respectively. The prevalence and maximum concentrations for these analytes likely reflect the large amount of data contributed by the studies of Hao et al. (2006) and Lissemore et al. (2006); these authors studied pharmaceutical concentrations in surface waters downstream from agricultural operations. The other antibiotics were detected at lower frequencies and at maximum concentrations less than 0.2 μg/L.

Prescription and non-prescription drugs were frequently detected in surface waters. The range of concentrations reported for each of the pharmaceutical compounds is illustrated in Fig. 4. Carbamazepine (antiepileptic), ibuprofen

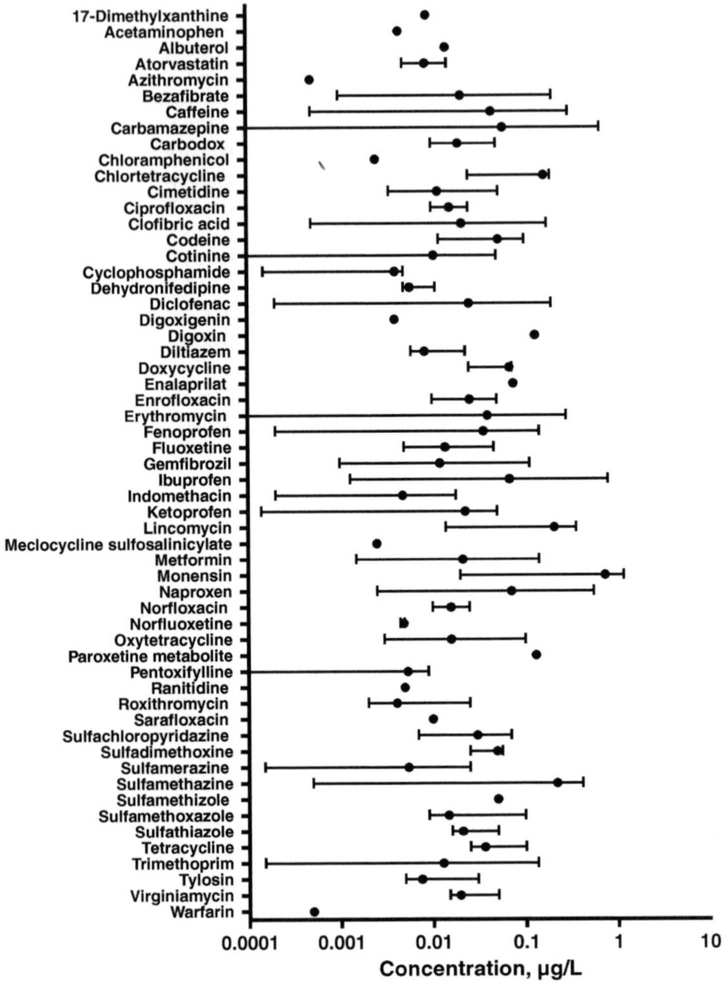


**Fig. 4** Minimum, weighted mean, and maximum concentrations reported for pharmaceutical compounds in surface waters

(analgesic/anti-inflammatory), and naproxen (analgesic/anti-inflammatory) were detected at the highest concentrations (0.55–0.79 μg/L); the frequency with which they were present in the samples ranged from 16 to 70% (Table 6). Caffeine and cotinine were among the substances most frequently detected (68–72%) and are generally considered biomarkers of human excretion. Of the prescription drugs, atorvastatin (lipid regulator), cimetidine (antacid), fenoprofen (anti-inflammatory), fluoxetine (anti-depressant), gemfibrozil (anti-hyperlipidemic), and pentoxifylline (vasodilator) were detected at frequencies greater than 25% with maximum concentrations ranging from 0.009 to 0.142 μg/L. Other prescription drugs (bezafibrate,

**Table 7** Summary statistics for pharmaceutical compounds in sediments

| Analyte | *n* | Freq det (%) | Median (ng/g dwt) | Min (ng/g dwt) | Max (ng/g dwt) |
|---|---|---|---|---|---|
| *Veterinary and human antibiotics* | | | | | |
| Azithromycin | 18 | 0 | 0.5 | 0.5 | 0.5 |
| Erythromycin | 18 | 11 | 4.4 | 0.5 | 8.2 |
| Sulfamethoxazole | 18 | 22 | 2.0 | 1.8 | 3.3 |
| Thiabendazole | 18 | 0 | 0.5 | 0.5 | 0.5 |
| Trimethoprim | 18 | 11 | 3.8 | 0.3 | 7.4 |
| *Prescription drugs* | | | | | |
| Albuterol | 18 | 0 | 0.5 | 0.5 | 0.5 |
| Carbamazepine | 18 | 0 | 0.5 | 0.5 | 0.5 |
| Cimetidine | 18 | 0 | 0.5 | 0.5 | 0.5 |
| Codeine | 18 | 0 | 0.5 | 0.5 | 0.5 |
| Dehydronifedipine | 18 | 6 | 12.0 | 12.0 | 12.0 |
| Diltiazem | 18 | 11 | 13.0 | 0.8 | 25.0 |
| Fluoxetine | 18 | 0 | 0.5 | 0.5 | 0.5 |
| Ranitidine | 18 | 0 | 0.5 | 0.5 | 0.5 |
| Warfarin | 18 | 0 | 0.5 | 0.5 | 0.5 |
| *Non-prescription drugs* | | | | | |
| Acetaminophen | 18 | 0 | 0.5 | 0.5 | 0.5 |
| Caffeine | 18 | 28 | 7.7 | 1.5 | 12.0 |
| Cotinine | 18 | 0 | 0.5 | 0.5 | 0.5 |
| 1,7-Dimethylxanthine | 18 | 0 | 0.5 | 0.5 | 0.5 |
| Diphenhydramine | 18 | 50 | 12 | 0.3 | 75.0 |
| Miconazole | 18 | 28 | 7.7 | 6.1 | 11.0 |

clofibric acid, cyclophosphamide, diclofenac, diltiazem, indomethacin, and metformin) were detected at frequencies less than 25%, and their maximum concentrations were generally less than 0.2 μg/L.

As summarized in Table 7, limited data are available for the concentrations of pharmaceutical compounds in sediments. Of the 20 compounds that have been analyzed, 8 were detected. Diphenhydramine was most frequently detected at median and maximum concentrations of 12 and 75 ng/g dwt, respectively. Caffeine, miconazole, and sulfamethoxazole were detected at frequencies greater than 20% at maximum concentrations ranging from 3.3 to 12 ng/g dwt. The other compounds were detected at lower frequencies and at maximum concentrations less than 25 ng/g dwt.

In summary, the available data suggest that pharmaceutical compounds may occur at nanogram to low microgram per liter concentrations in surface water samples. At present, there are no standards, guidelines, or criteria with which to compare the concentrations. Pharmaceutical compounds are most frequently detected near the point of discharge from wastewater treatment plants or agricultural operations, especially in locations where hydrologic conditions result in minimal dilution. Collectively, the frequency with which pharmaceutical compounds have been detected in surface waters of the Great Lakes Basin is low; detectable concentrations have been reported for 34% of the 980 data points. However, as noted by Metcalfe et al. (2003), detectable concentrations of some drugs, including clofibric acid, ketoprofen, fenoprofen, and carbamazepine, may be present at low nanogram per liter levels in surface waters taken from open waters or from sites that are relatively remote from the influence of wastewater discharges.

## *3.3 Organic Wastewater Contaminants, Hormones, and Steroids*

A variety of chemicals have been detected in surface waters downstream from wastewater treatment plants. The organic wastewater contaminants discussed in this section include a variety of industrial chemicals and personal care products not addressed in other sections of this report. Available information on concentrations of human and animal hormones and steroids is also addressed.

Much of the available information on the concentration and distribution of organic wastewater contaminants in the Great Lakes Basin were from Kolpin et al. (2002). As previously noted, surface water samples were collected during 1999 and 2000 as part of a nationwide reconnaissance program from a network of 139 streams across 30 states, of which 19 sites were located within the Lake Michigan drainage basin (sampling locations not illustrated). The selection of the sampling sites was biased toward areas considered to be susceptible to contamination from human, industrial, and agricultural wastewaters. Of the 95 organic wastewater contaminants included in the study, 40 represented a variety of industrial chemicals (anti-corrosives, antioxidants, plastics monomers, plasticizers, solvents, etc.), personal care products (deodorizers, fragrances, disinfectants, insect repellant), polycyclic aromatic hydrocarbons (PAHs), as well as human and animal hormones and steroids.

Tertuliana et al. (2008) examined the occurrence and distribution of a broad range of pharmaceutical and organic wastewater compounds in the Tinkers Creek watershed in northeastern Ohio. In addition to using passive samplers, sediment samples were collected at each of the sampling stations. The samples were collected during May–June of 2006 from 18 locations situated upstream and downstream from wastewater treatment plant discharges to the mainstream and tributaries of Tinkers Creek. Of the 57 organic wastewater compounds included in the analysis, a total of 33 organic wastewater compounds and 4 steroids were detected in sediment at one or more of the sites.

Based on the available literature, a database was developed containing 336 reported values for the 66 compounds analyzed in surface waters and sediments from the Great Lakes Basin. As summarized in Table 8, between 4 and 29 surface water samples were analyzed for each of the contaminants. The frequencies with which the various compounds were detected in water ranged from 0 to as high as 90%. Reported detection limits for the compounds ranged from 0.005 to 2.0 μg/L. The concentrations of organic wastewater contaminants detected in sediment are

**Table 8** Summary statistics for organic wastewater constituents in surface waters

| Analyte | *n* | Freq det (%) | Min (μg/L) | Max (μg/L) |
|---|---|---|---|---|
| *Anti-corrosives/antioxidants* | | | | |
| Butylated hydroxy toluene | 7 | 14.3 | < 0.08 | 0.1 |
| 3-*tert*-Butyl-4-hydroxy anisole | 7 | 0.0 | < 0.12 | |
| 2,6-Di-*tert*-*p*-benzoquinone | 7 | 14.3 | < 0.07 | 0.06 |
| 2,6-Di-*tert*-butyl-phenol | 7 | 14.3 | < 0.08 | 0.1 |
| 5-Methyl-1*H*-benzotriazole | 4 | 75.0 | < 0.15 | 0.55 |
| *Deodorizer/fragrances* | | | | |
| Acetophenone | 7 | 14.3 | < 0.1 | 0.1 |
| 1,4-Dichlorobenzene | 7 | 28.6 | < 0.03 | 0.17 |
| *Disinfectants* | | | | |
| 4-Methyl phenol | 7 | 42.9 | < 0.04 | 0.07 |
| Phenol | 7 | 28.6 | < 0.08 | 0.6 |
| Triclosan | 29 | 89.7 | < 0.004 | 0.3 |
| *Insect repellant* | | | | |
| *N*,*N*-Diethyltoluamide | 4 | 75.0 | < 0.08 | 0.24 |
| *Plasticizers/plastics manufacture* | | | | |
| Bis(2-ethylhexyl)adipate | 7 | 0.0 | < 1.5 | |
| Bis(2-ethylhexyl)phthalate | 7 | 14.3 | < 2.0 | 20 |
| Bisphenol A | 7 | 57.1 | < 0.09 | 0.8 |
| Diethyl phthalate | 4 | 25.0 | < 0.25 | 0.2 |
| Ethanol, 2-butoxy phosphate | 7 | 42.9 | < 0.07 | 0.82 |
| Phthalic anhydride | 7 | 42.9 | < 0.15 | 0.9 |
| Triphenyl phosphate | 7 | 0.0 | < 0.1 | |
| *Polycyclic aromatic hydrocarbons* | | | | |
| Anthracene | 7 | 14.3 | < 0.05 | 0.05 |
| Benzo(*a*)pyrene | 7 | 28.6 | < 0.05 | 0.11 |
| Fluoranthene | 7 | 42.9 | < 0.03 | 0.2 |
| Naphthalene | 7 | 14.3 | < 0.02 | 0.02 |
| Phenanthrene | 7 | 42.9 | < 0.05 | 0.06 |
| Pyrene | 7 | 57.1 | < 0.03 | 0.27 |
| *Solvents* | | | | |
| Tetrachloroethylene | 7 | 14.3 | < 0.03 | 0.2 |
| *Steroids/hormones* | | | | |
| *cis*-Androsterone | 5 | 20.0 | < 0.005 | 0.031 |
| Cholesterol | 12 | 66.7 | < 1.5 | 2.38 |
| Coprostanol | 12 | 50.0 | < 0.6 | 0.73 |
| Equilenin | 5 | 0.0 | < 0.005 | |
| Equilin | 5 | 0.0 | < 0.005 | |
| 17α-Estradiol | 5 | 20.0 | < 0.005 | 0.001 |
| 17α-Ethynyl estradiol | 5 | 0.0 | < 0.005 | |
| 17β-Estradiol | 12 | 0.0 | < 0.005 | |
| Estriol | 5 | 20.0 | < 0.005 | 0.02 |
| Estrone | 5 | 0.0 | < 0.005 | |

**Table 8** (continued)

| Analyte | $n$ | Freq det (%) | Min (μg/L) | Max (μg/L) |
|---|---|---|---|---|
| Mestranol | 5 | 0.0 | < 0.005 | |
| 19-Norethisterone | 5 | 0.0 | < 0.005 | |
| Progesterone | 5 | 0.0 | < 0.005 | |
| Stigmastanol | 4 | 0.0 | < 2.0 | |
| Testosterone | 5 | 0.0 | < 0.005 | |

summarized in Table 9. The frequencies with which the various compounds were detected varied from 0 to 100%. As noted in the table, between 5 and 73 sediment samples were analyzed for the various contaminants. When available, the concentrations reported in Great Lakes Basin were compared to US Environmental Protection Agency (US EPA), Environment Canada (EC), and European Union (EU) regulatory criteria and guidelines (Table 2).

**Table 9** Summary statistics for organic wastewater constituents in sediments

| Analyte | $n$ | Freq det (%) | Median (ng/g dwt) | Min (ng/g dwt) | Max (ng/g dwt) |
|---|---|---|---|---|---|
| *Anti-corrosives/antioxidants* | | | | | |
| 3-*tert*-Butyl-4-hydroxy anisole | 18 | 6 | 12.5 | 12.5 | 12.5 |
| *Deodorizer/fragrances* | | | | | |
| Acetophenone | 18 | 89 | 35 | 10 | 90 |
| AHTN | 18 | 56 | 30 | 10 | 80 |
| Camphor | 18 | 0 | 12.5 | 12.5 | 12.5 |
| 1,4-Dichlorobenzene | 18 | 17 | 16 | 16 | 16 |
| HHCB | 18 | 50 | 60 | 20 | 390 |
| Indole | 18 | 100 | 55 | 30 | 1, 100 |
| Isoborneol | 18 | 0 | 12.5 | 12.5 | 12.5 |
| Isoquinoline | 18 | 0 | 12.5 | 12.5 | 12.5 |
| D-Limolene | 18 | 11 | 10 | 10 | 10 |
| Menthol | 18 | 0 | 12.5 | 12.5 | 12.5 |
| 3-Methyl indole | 18 | 100 | 30 | 10 | 150 |
| *Disinfectants/preservatives* | | | | | |
| 4-Methyl phenol | 18 | 89 | 35 | 10 | 260 |
| Phenol | 18 | 83 | 70 | 20 | 220 |
| Triclosan | 18 | 17 | 36 | 30 | 56 |
| *Insect repellant* | | | | | |
| *N,N*-Diethyltoluamide | 18 | 0 | 12.5 | 12.5 | 12.5 |
| *Chemicals/plastics manufacture* | | | | | |
| Anthraquinone | 18 | 100 | 70 | 20 | 240 |
| Benzophenone | 18 | 0 | 12.5 | 12.5 | 12.5 |
| Benzylbutylphthalate | 5 | 0 | 150 | 150 | 150 |
| Bis (2-ethylhexyl)phthalate | 23 | 97 | 4, 030 | 30 | 29, 700 |
| Bis(isononyl)phthalate | 5 | 0 | 150 | 150 | 150 |

**Table 9** (continued)

| Analyte | *n* | Freq det (%) | Median (ng/g dwt) | Min (ng/g dwt) | Max (ng/g dwt) |
|---|---|---|---|---|---|
| Bisphenol A | 73 | 38 | 21.6 | 0.075 | 60 |
| Carbazole | 18 | 100 | 50 | 20 | 130 |
| 4-Cumylphenol | 18 | 0 | 12.5 | 12.5 | 12.5 |
| Dibutyl phthalate | 5 | 0 | 150 | 150 | 150 |
| Diethyl phthalate | 33 | 11 | 80 | 10 | 150 |
| Isopropylbenzene | 18 | 6 | 12.5 | 12.5 | 12.5 |
| 2,2,4,4-Tetrabromodiphenylether | 18 | 0 | 12.5 | 12.5 | 12.5 |
| Tributylphosphate | 18 | 0 | 12.5 | 12.5 | 12.5 |
| Triphenyl phosphate | 18 | 0 | 12.5 | 12.5 | 12.5 |
| Tris(2-butoxyethyl) phosphate | 18 | 39 | 30 | 20 | 70 |
| Tris(dichloroisopropyl)phosphate | 18 | 0 | 12.5 | 12.5 | 12.5 |
| *Polycyclic aromatic hydrocarbons* | | | | | |
| Anthracene | 18 | 94 | 50 | 10 | 110 |
| Benzo(*a*)pyrene | 18 | 100 | 140 | 30 | 390 |
| 2,6-Dimethylnaphthalene | 18 | 100 | 25 | 10 | 60 |
| Fluoranthene | 18 | 100 | 540 | 80 | 1,400 |
| 1-Methylnaphthalene | 18 | 89 | 20 | 10 | 40 |
| 2-Methylnaphthalene | 18 | 100 | 20 | 10 | 50 |
| Naphthalene | 18 | 100 | 20 | 10 | 40 |
| Phenanthrene | 18 | 94 | 250 | 60 | 720 |
| Pyrene | 18 | 100 | 400 | 60 | 1,100 |
| *Solvents* | | | | | |
| Isophorone | 18 | 0 | 12.5 | 12.5 | 12.5 |
| *Steroids/hormones/natural products* | | | | | |
| Cholesterol | 18 | 100 | 560 | 260 | 2,800 |
| Coprostanol | 18 | 83 | 80 | 20 | 300 |
| β-Sitosterol | 18 | 100 | 620 | 210 | 2,500 |
| β-Stigmastanol | 18 | 83 | 140 | 30 | 1,100 |

A variety of industrial chemicals (anti-corrosives/antioxidants, substances used in chemicals or plastics manufacture, solvents, etc.) were detected in both surface waters and sediments. Of the industrial chemicals detected in water (Table 8), the anti-corrosive agent, 5-methyl-1*H*-benzotriazole was detected in 75% of the samples at concentrations as high as 0.55 μg/L. This frequency of detection was higher than found in the nationwide survey (31.5%; Kolpin et al. 2002), although a higher maximum concentration (2.4 μg/L) was reported elsewhere in the United States. Several compounds related to plastics manufacture, including the monomers bisphenol A (BPA) and phthalic anhydride, and a plasticizer (ethanol-2-butoxy phosphate), were also among those xenobiotics that were frequently detected in Great Lakes surface waters. The other chemicals, including antioxidants, plasticizers, and solvents, were detected less frequently, ranging from their absence to being present in 25% of the samples. The concentrations of most of the industrial chemicals in water were low

(<1 μg/L), with the exception of bis(2-ethylhexyl)phthalate (DEHP), which was detected in a single sample at 20 μg/L. This value exceeds the EC Interim Water Quality Guideline (16.0 μg/L) and the EU predicted no effect value (1.3 μg/L).

Of the industrial chemicals detected in sediment (Table 9), bis(2-ethylhexyl)phthalate was detected in 97% of the samples at median and maximum concentrations of 4,030 and 29,700 ng/g dwt, respectively. The highest concentration, which exceeds the EU PNEC (predicted no-effect concentration) for sediment (21,500 ng/g dwt), was reported by McDowell and Metcalfe (2001) in samples taken in the vicinity of the wastewater treatment plant discharge in Hamilton Harbor, ON. In contrast, bis(2-ethylhexyl)phthalate concentrations detected in Tinkers Creek downstream from wastewater discharges ranged from 30 to 120 ng/g dwt (Tertuliana et al. 2008). Anthraquinone and carbazole were also frequently detected (100%) in sediment at concentrations ranging from 20 to 240 ng/g dwt. The other chemicals, including various antioxidants, plasticizers, and solvents, were detected less frequently and at concentrations less than 100 ng/g dwt.

BPA, a monomer used in the manufacture of epoxy resins and polycarbonate plastics, was detected in 57.1% of the surface water samples at concentrations ranging from below the detection limit to as high as 0.8 μg/L (Table 8). These results are somewhat higher than those of a recent statistical analysis of BPA exposures in North American aquatic environments (Klečka et al. 2009). Based on the analysis of 1,068 weighted observations for North American surface waters, BPA has been reported at concentrations above the detection limit in 20% of analyzed samples. Median and 95th percentile concentrations for North American waters were 0.081 μg/L and 0.47 μg/L, respectively. As summarized in Table 9, BPA has also been detected in 38% of the sediment samples at concentrations ranging from below the detection limit to 60 ng/g dwt. Chu et al. (2005) reported that BPA was detected in 65% of sediment samples from Lake Erie ($n = 55$) at concentrations ranging up to 6.1 ng/g dwt. Tertuliana et al. (2008) detected BPA in 11% of the samples from Tinkers Creek, with estimated concentrations ranging from 20 to 60 ng/g dwt.

Risk assessments for BPA have been conducted by several regulatory authorities around the world (Advanced Industrial Science and Technology (AIST) 2007; Environment Canada and Health Canada (EC and HC) 2008; European Commission (EUC) 2008a). The European Chemicals Bureau of the EUC (EUC 2008a) has recently completed a comprehensive risk assessment for BPA and has defined PNECs for water and sediment of 1.5 μg/L and 63 μg/kg dwt, respectively. The PNEC for water was derived using a species sensitivity distribution approach, based on chronic data for 16 different species across a wide range of trophic levels. Environment Canada (EC and HC 2008) has also recently finalized a screening assessment for BPA. The PNEC values for water and sediment are 0.175 μg/L and 0.01 mg/L, respectively. Note that the PNEC value for sediment was derived from a water-only exposure study. Considering the available data for BPA in the Great Lakes, the maximum concentrations for water and sediment are below the European PNEC values. However, the maximum concentration exceeds the Canadian PNEC for water, but is below the PNEC for sediment organisms.

Among the personal care products (deodorizers/fragrances, disinfectants, insect repellants, etc.), triclosan and *N*,*N*-diethyltoluamide were among those most frequently detected (>75%) in Great Lakes waters (Table 8). The substances 1,4-dichlorobenzene, phenol, and 4-methylphenol were detected less frequently and at concentrations below 1 μg/L. Many of these substances were also present in sediment (Table 9). Among those most frequently detected were a number of natural products used commercially, including acetophenone, indole, 3-methyl indole, 4-methyl phenol, and phenol; concentrations in sediment ranged from 10 to 1,100 ng/g dwt. The maximum concentrations for 1,4-dichlorobenzene and phenol in water and sediment are below EC and EU criteria (Table 2).

Triclosan was frequently detected in the surface water samples from the Great Lakes Basin (89.7%) and had concentrations ranging from below the detection limit to as high as 0.3 μg/L. In contrast, triclosan was infrequently detected (17%) in sediment and had maximum concentrations less than 100 ng/g dwt. Triclosan is an antimicrobial disinfectant agent used for over 30 years in a wide array of personal care and consumer products. Triclosan was detected less frequently in water in the nationwide reconnaissance study (57.6%; Kolpin et al. 2002), although the maximum reported concentration was higher (2.3 μg/L) elsewhere in the United States. Hua et al. (2005) reported low triclosan concentrations, ranging from 4 to 8 ng/L in surface water samples taken from two sites along the Detroit River shoreline, downstream of the Little River Wastewater Treatment Plant (Windsor, ON). Servos et al. (2007) reported low levels of triclosan (34 ng/L) in river water samples collected in southern Ontario.

Valters et al. (2005) reported the presence of triclosan and methyl-triclosan in the blood plasma of pelagic and benthic fish (13 species) that were collected from the Detroit River during 2001 and 2002. Concentrations of triclosan in plasma of pelagic fish ranged from 1.87 to 10.27 ng/g wwt. Similar concentrations were reported in plasma from benthic fish (0.76–5.53 ng/g wwt). The levels of methyl-triclosan were lower, ranging from 0.0004 to 0.013 ng/g wwt. The source of methyl-triclosan was attributed to biological methylation during wastewater treatment (Balmer et al. 2004; Lindstrom et al. 2002). The authors noted that anthropogenic sources of triclosan in the Detroit River system contribute to the concentrations detected in the fish; however, when compared to the levels of other organohalogen compounds measured in the plasma, the concentrations of triclosan were comparable to those for total polybrominated diphenyl ethers (PBDEs; 0.16–21.07 ng/g wwt), but considerably lower than the total PCB concentrations detected in the same fish (10.4–909.1 ng/g wwt).

The insect repellant *N*,*N*-diethyltoluamide (DEET) was frequently detected (75%) in water but not in sediment samples from the Great Lakes Basin. According to the US EPA (1998), approximately 30% of the US population uses DEET annually as an insect repellant. With the spread of the West Nile virus, there have been frequent official recommendations that residents in affected areas use products containing DEET. Concentrations in the Great Lakes Basin ranged from below the detection limit (0.08 μg/L) to as high as 0.24 μg/L. These results are consistent with those of Sandstrom et al. (2005) who reported that DEET was detected at levels of 0.02 μg/L or greater in 73% of the 56 stream sites sampled in the United States during the year 2000. Although DEET was frequently detected at all sites, the median concentration was low (0.05 μg/L). The highest concentrations (maximum of 1.1 μg/L) were found in samples collected from urban areas.

A variety of PAHs were also measured in water and sediment samples from the Great Lakes Basin (Tables 8 and 9). Combustion by-products, including pyrene, phenanthrene, fluoranthene, and benzo(a) pyrene, were among the most frequently detected compounds, and they displayed maximum concentrations ranging from 0.06 to 0.27 μg/L in water and from 390 to 1,400 ng/g dwt in sediment. Although anthracene and naphthalene were detected less frequently in water, they were generally found in all sediment samples. Several other PAHs that are components of petroleum hydrocarbon fuels (gasoline, diesel fuel) were also detected in sediments. When the maximum concentrations are compared to regulatory criteria developed by the EC and EU, naphthalene and phenanthrene concentrations in water are below the guidelines for water and sediment, but many of the other PAH compounds exceed those limits (Table 2).

Lopes and Furlong (2001) reported the occurrence of 65 semi-volatile organic contaminants (SVOCs) in streambed samples collected between 1992 and 1995 from 536 sites representing 20 major river basins. Although the details of the investigation were not available for our study, it is apparent that some of the samples were collected in eastern Wisconsin from the Western Lake Michigan drainage basin. From the summary information provided in the paper, the highest SVOC concentrations, which included a variety of PAHs, phthalates, and phenols, were detected in samples collected from the northeastern United States and Great Lakes Basins. PAH and phthalate concentrations were 10 times higher in areas influenced by urban activities (high population densities and large metropolitan areas) than at sites with other land uses. The highest total PAH (170,000 μg/kg) and total phenol (4,900 μg/kg) concentrations were measured near Milwaukee, WI. The authors reported that on the basis of sediment quality guidelines, adverse effects are probable at 7.5% and are possible at 16.2% of the sites. High PAH concentrations contributed the greatest potential for adverse effects on aquatic ecosystems.

A variety of human, animal, and plant hormones and steroids were also detected in water and sediment samples from the Great Lakes Basin (Tables 8 and 9). Cholesterol (plant and animal steroid) and coprostanol (fecal steroid) were present in more than half of the water, and nearly all of the sediment samples. Two plant steroids (sitosterol and stigmastanol) were also frequently detected in sediment. *cis*-Androsterone (urinary steroid), 17α-estradiol (hormone), and estriol (hormone) were also detected in 20% of the water samples. Detection frequencies for the latter compounds, in waters of the Great Lakes, are comparable to those reported by Kolpin et al. (2002), although generally higher concentrations were found elsewhere in the United States. Higher detection frequencies and maximum concentrations were also reported by Kolpin et al. (2002) for many other hormones and steroids that were not detected in the Great Lakes Basin. For example, 17α-ethynyl estradiol and 17β-estradiol were present in 10–16% of the samples nationwide at maximum concentrations from 0.009 to 0.073 μg/L. As noted by Kolpin et al. (2002), when potency and aquatic toxicity are considered, measured concentrations of reproductive hormones may have greater implications for the health of aquatic organisms than measured concentrations of the other emerging contaminants. Previous research (Baronti et al. 2000; Purdom et al. 1994; Routledge et al. 1998) has shown that even low-level exposure (<0.001 μg/L) to select hormones can illicit adverse effects in aquatic species.

In summary, a variety of organic wastewater contaminants have been detected in the Great Lakes Basin, representing a wide range of residential, industrial, and agricultural origins. Of those addressed here, the most frequently detected in water included 5-methyl-1*H*-benzotriazole, triclosan, *N*,*N*-diethyltoluamide, BPA, and several PAHs. Measured concentrations for Great Lakes waters were generally low (<1 μg/L) and generally were below regulatory guidelines or criteria, other than as noted above. Many of the compounds, however, do not have such guidelines established. Concentrations of many other organic wastewater contaminants that were detected in the nationwide reconnaissance, including current-use pesticides, pharmaceuticals, alkylphenol ethoxylates, synthetic musks, and flame retardants, are discussed in other sections of this report.

## *3.4 Alkylphenol Ethoxylates*

Alkylphenol ethoxylates (APEO) have been used as surfactants for more than 50 years. APEO in use today include nonylphenol ethoxylates (NPEO), comprising about 80% of the market, whereas octylphenol ethoxylates (OPEO) comprise about 20% of the market. These compounds have been the subject of considerable regulatory attention, primarily from concerns about their aquatic toxicity and weak endocrine activity. Risk assessments have been conducted by regulatory authorities around the world (EC and HC 2001; EUC 2002a; United Kingdom Environment Agency; UK EA 2005; US EPA 1996), and risk management strategies have been implemented, which include their use reduction and/or elimination in certain applications, along with voluntary reductions in their use in several countries. Today, NPEO are primarily used for industrial applications that include pulp and paper production, textile manufacturing, and use in the formulation of crop protection chemicals. They are also used in industrial and institutional cleaners and detergents. Although nonylphenol (NP) is used primarily in the manufacture of NPEO, some is also used in the production of resins and plastics stabilizers. In contrast, octylphenol (OP) is produced in significantly lower volumes and is used primarily as a chemical intermediate in the production of phenolic resins. Because of their use patterns, most of the spent APEO are discharged into wastewater treatment systems, which have been shown to be effective in reducing loads entering the environment (Melcer et al. 2007). Although most APEO are removed during wastewater treatment, metabolites consisting of alkylphenols (AP), low mole APEO, and alkylphenol ether carboxylates (APEC) have been reported in effluents, streams, and rivers across North America.

Concentrations of APEO and their metabolites have been reported in environmental samples by numerous investigators over the past 15 years. As discussed below, the concentrations of APEO in the Great Lakes Basin, including levels in surface water, sediment, and biota, have been reported in 10 studies. The geographic distribution of the sampling sites is shown in Fig. 5a and b.

Bennie et al. (1997) reported on the occurrence of AP and low mole APEO in waters of the Great Lakes Basin and the upper St. Lawrence River. Surface water samples collected during 1994 and 1995 from 35 sites showed measurable quantities of NP and OP in 24% of the samples at concentrations ranging from below the detection limit to 0.92 μg/L. Low mole NP1EO and NP2EO were present in 32–58% of the water samples at levels as high as 10 μg/L. Sediment samples from all nine heavily industrialized sites had detectable levels of NP at concentrations ranging from 0.17 to 72 μg/g dwt. OP was detected in 89% of the samples at lower concentrations (<0.01–1.8 μg/g dwt). NP1EO and NP2EO were present in 66% of the samples; the maximum concentrations ranged from 6 to 38 μg/g dwt. The authors noted that the highest concentrations were all below acute toxicity thresholds, but were sufficient to raise concern with regard to long-term effects on fish.

Bennett and Metcalfe (1998) analyzed for NP and OP in 28 sediment samples collected from both industrialized and pristine regions throughout Lakes Huron, Erie, and Ontario. The results indicated that the AP are widely distributed in sediments from the lower Great Lakes. NP and OP concentrations were generally low (<1 μg/g dwt) except in sediments collected near urban and industrial centers. In the vicinity of wastewater treatment discharges, NP and OP concentrations as high as 37.8 μg/g dwt and 23.7 μg/g dwt were reported. In a follow-up study, Bennett and Metcalfe (2000) examined the spatial distribution of APEO and their degradation products in receiving waters near sewage treatment plant discharges in Hamilton Harbor and in the Detroit River. The investigators collected sediment samples, deployed semi-permeable membrane devices, and caged freshwater mussels at 24 locations in the summer of 1996. NP concentrations were elevated (>20 μg/g dwt) in the Detroit River downstream of both the Windsor and Detroit wastewater treatment plant discharges, but rapidly declined to less than 1 μg/g dwt downstream. Among the reported NP concentrations in Great Lakes sediments, the highest (110 μg/g) was that detected in the vicinity of the outflow of the Burlington wastewater treatment plant in Hamilton Harbor. Analysis of the spatial data for Hamilton Harbor indicated that the concentrations of NP and the low mole NPEO decreased by a factor of 2 over distances of 30–50 m. The authors concluded that the environmental distribution of APEO is expected to be localized in areas close to the point of discharge and are likely to decline at sites a few hundred meters from the source.

Kannan et al. (2001) analyzed sediments from the upper Detroit and lower Rouge Rivers in Detroit, MI as well as samples from a non-point source location in Lake Michigan (near Muskegon, MI). Several industrial facilities and combined sewer overflows located along the two rivers have historically impacted the levels of sediment contaminants. Thirty four samples were collected in June 1998 and analyzed for NP, OP, and a variety of chlorinated aromatic compounds. The distribution of

A. Surface waters

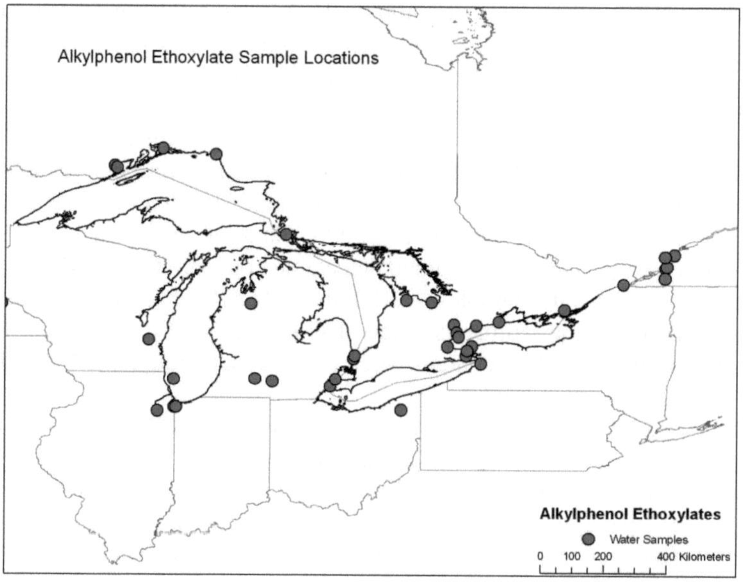


B. Sediments

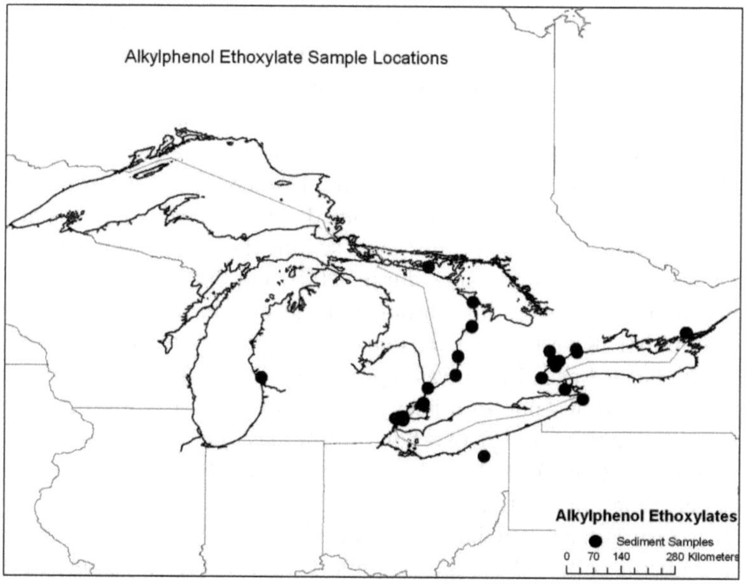


**Fig. 5** Sampling locations for alkylphenol ethoxylates. **a** Surface waters and **b** sediments

NP and OP in the rivers suggested the presence of localized, but multiple, sources of contamination for each class of compounds. NP and OP were detected in 70% and 65% of the sediment samples with concentrations ranging from undetectable to as high as 60,000 and 4,050 μg/kg, respectively. Neither compound was detected in sediment samples collected from Lake Michigan.

Kannan et al. (2003) also conducted a survey of water, sediment, and fish samples from the Kalamazoo River. Samples were collected along transverse transects both upstream and downstream from wastewater discharges located in Battle Creek and Kalamazoo, MI, during July and August, 2000. Three of 36 sediment samples and 1 of 24 surface water samples contained detectable concentrations. NP and NP1EO concentrations in the water sample were 1.1 and 2.0 μg/L, respectively; measurable NP concentrations in the sediments ranged from 5.8 to 15.3 μg/kg dwt. Of the 10 fish caught during the study, 1 contained detectable concentrations of NP (3.4 μg/kg wwt).

Keith et al. (2001) reported the results of analysis of 183 fish (representing 8 species) collected in 1999 from the Kalamazoo River and Lake Michigan. NP concentrations were detectable in 41% of the samples; measured concentrations ranged from 3.3 to 29.1 μg/kg wwt. Traces of NP1EO were present in 11% of the fish, but the levels were not quantifiable. Higher mole NPEO were not detected.

Kolpin et al. (2002) analyzed for NP and low mole NPEO and OPEO in nine water samples from the Great Lakes Basin during the US nationwide reconnaissance study (1999–2000). NP was detected in 55% of the samples; measured concentrations ranged from 0.25 to 3 μg/L. NP1EO and NP2EO were detected less frequently (11%) with maximum concentrations of 1 μg/L. OP and low mole OPEO were detected in some samples at lower concentrations.

Mayer et al. (2007) analyzed for the presence of AP in the coastal marsh, Cootes Paradise, ON, Canada. This coastal wetland and nature sanctuary, located west of Hamilton Harbor, receives discharges from the Dundas, ON, wastewater treatment plant as well as inputs from several combined sewer overflows from the city of Hamilton, ON. Samples of water ($n = 6$) and sediment ($n = 21$) were collected from both open water and shoreline sites during 2001 and 2002. NP, OP, and NP1EO concentrations in water were low throughout the marsh, ranging from below the detection limit to 0.28 μg/L. However, one of the highest of the NP3–17EO concentrations in the Great Lakes Basin was reported (91.7 μg/L). The authors attributed this high concentration value to a heavy rainfall event that produced a large discharge from a combined sewer overflow on the day the samples were collected. Sediment concentrations for NP and NPEO were generally consistent, ranging from below the detection limit to 1.75 μg/g dwt.

Rice et al. (2003) examined the occurrence of a variety of AP and APEO in water, sediment, and fish collected along a 74-mile length of the Cuyahoga River, OH. Water and sediment samples, as well as 11 or 12 carp were collected from 8 sites along the river in the summer of 2000. Maximum AP/APEO concentrations in water and fish were observed downstream of the Akron wastewater treatment plant. NP concentrations in the water ranged from 0.1 to 0.5 μg/L; total NP0–3EO concentrations were slightly higher (0.13–5.1 μg/L). Total OP0–3EO concentrations were low (0.005–0.19 μg/L). Average total NP0–5EO concentrations in the fish ranged from

32 μg/kg at the upstream (clean) site to as high as 920 μg/kg; NP1EO was the predominant homolog detected. NP concentrations in fish varied from 7 to 110 μg/kg. According to the authors, the NP levels could be considered moderate when compared to maximum NP residues that have been reported in the literature. However, NP1EO and NP2EO concentrations were among the highest reported for fish in US waters. NP concentrations in the sediment ranged from 75 to 340 μg/kg, with total NP0–4EO ranging from 250 to 1,020 μg/kg. Total OP0–5EO levels in the sediment ranged from 20 to 74 μg/kg. Although AP/APEO concentrations increased below the wastewater discharge in Akron, OH, the highest concentrations were present at the most downstream site near Cleveland.

Sabik et al. (2003) examined the prevalence of APEO in surface water, sediment, and caged mussels from the St. Lawrence River at sites both upstream and downstream from the Montreal, QC, effluent outfall. At the time of the study (1999), the facility was the largest primary physico-chemical treatment plant in North America. In contrast to many of the other studies, the target analytes included a broad range of NPEO homologs (NP1EO–NP16EO) as well as NP, OP, and the ether carboxylates. The higher mole NPEO were predominant in the surface waters and were present in samples ($n = 4$) collected both upstream and downstream of the discharge. Concentrations of NP>7EO ranged from below the detection limit to as high as 10.3 μg/L. NP, OP, and the low mole NPEO were generally below the detection limit. In contrast, NP and the low mole NPEO were detected in the sediments ($n = 6$), with the highest concentrations occurring downstream of the discharge. To assess the potential for bioconcentration of APEO, mussels taken from a reference lake were caged and submerged for 62 days upstream and downstream of the facility. The authors noted a slight, but not significant bioconcentration of NP5EO–NP8EO in the mussels, which was more noticeable at the downstream site.

As presented in Table 10a, concentrations of APEO and their metabolites have been reported for 87 surface water samples obtained during the years from 1994 to 2002 at 43 locations throughout the Great Lakes Basin. As previously discussed, many of the sampling locations were downstream from the discharges of wastewater treatment plants. Although concentrations have been reported for as many as 24 different analytes, to facilitate interpretation, the analytes were first grouped into categories on the basis of structure and data availability, as follows: alkylphenols (NP, OP), low mole ethoxylates (NP1EO, NP2EO), higher mole ethoxylates (NP3–17EO, OPEO), and ether carboxylates (nonylphenol ether carboxylates (NPEC), octylphenol ether carboxylates (OPEC)). These groupings are intended to represent the majority of compounds found across all of the studies.

Of the 385 values reported for surface water, 46% were above the detection limits. NPEC were among the analytes detected most often in the samples (75%) with average concentrations of 0.97 ± 2.12 μg/L, which is consistent with the fact that ether carboxylates have been reported as the dominant biodegradation products of wastewater treatment (Melcer et al. 2007). Low and high mole NPEO were detected in 34–60% of the samples, with average concentrations ranging from 0.64 ± 1.22 to 2.01 ± 10.0 μg/L, respectively. The fact that high mole NPEO were detected at such levels is unusual. Further examination of the data indicated that samples were only analyzed for high mole NPEO in two studies (Mayer et al. 2007; Sabik et al. 2003).

**Table 10** Summary statistics for alkylphenol ethoxylates

| | *n* | Freq det (%) | Mean (μg/L) | SD (μg/L) | Median (μg/L) | 95th percentile (μg/L) | Min (μg/L) | Max (μg/L) |
|---|---|---|---|---|---|---|---|---|
| *A. Surface water* | | | | | | | | |
| NP | 69 | 36 | 0.25 | 0.49 | 0.005 | 1.30 | 0.0005 | 3.00 |
| NP1EO | 69 | 54 | 0.64 | 1.22 | 0.159 | 2.18 | 0.0010 | 7.80 |
| NP2EO | 68 | 34 | 1.08 | 2.53 | 0.015 | 7.80 | 0.0010 | 11.00 |
| NP3–17EO | 87 | 60 | 2.01 | 10.00 | 0.063 | 8.90 | 0.0005 | 91.70 |
| NPEC | 20 | 75 | 0.97 | 2.12 | 0.395 | 2.85 | 0.0010 | 9.67 |
| OP | 48 | 25 | 0.02 | 0.07 | 0.003 | 0.07 | 0.0005 | 0.47 |
| OPEO | 14 | 50 | 0.11 | 0.06 | 0.100 | 0.24 | 0.0500 | 0.30 |
| OPEC | 10 | 60 | 0.06 | 0.12 | 0.027 | 0.25 | 0.0010 | 0.41 |
| All alkylphenol ethxoyaltes (APE) | 385 | 46 | 0.86 | 4.95 | 0.045 | 3.81 | 0.0005 | 91.70 |

| | *n* | Freq det (%) | Mean (μg/g dwt) | SD (μg/g dwt) | Median (μg/g dwt) | 95th percentile (μg/g dwt) | Min (μg/g dwt) | Max (μg/g dwt) |
|---|---|---|---|---|---|---|---|---|
| *B. Sediment* | | | | | | | | |
| NP | 113 | 81 | 5.60 | 14.94 | 0.30 | 30.00 | 0.003 | 110.00 |
| NP1EO | 51 | 82 | 5.22 | 13.13 | 0.28 | 32.50 | 0.005 | 70.00 |
| NP2EO | 51 | 69 | 0.76 | 2.91 | 0.12 | 2.50 | 0.001 | 20.00 |
| NP3–17EO | 33 | 76 | 0.62 | 0.94 | 0.09 | 2.88 | 0.015 | 2.88 |
| NPEC | 16 | 13 | 0.20 | 0.12 | 0.27 | 0.27 | 0.001 | 0.27 |
| OP | 76 | 71 | 0.56 | 2.75 | 0.02 | 1.44 | 0.001 | 23.70 |
| OPEO | 7 | 100 | 0.05 | 0.02 | 0.06 | 0.07 | 0.020 | 0.07 |
| OPEC | 14 | 0 | 0.02 | 0.01 | 0.02 | 0.02 | 0.003 | 0.02 |
| All APE | 361 | 71 | 2.78 | 10.12 | 0.12 | 20.00 | 0.001 | 110.00 |

**Table 10** (continued)

| | n | Freq det (%) | Mean (μg/g wwt) | SD (μg/g wwt) | Median (μg/g wwt) | 95th percentile (μg/g wwt) | Min (μg/g wwt) | Max (μg/g wwt) |
|---|---|---|---|---|---|---|---|---|
| *C. Biota* | | | | | | | | |
| Fish | | | | | | | | |
| NP | 284 | 63 | 0.025 | 0.035 | 0.007 | 0.102 | 0.002 | 0.11 |
| NP1EO | 284 | 29 | 0.095 | 0.157 | 0.008 | 0.400 | 0.008 | 0.55 |
| NP2EO | 284 | 29 | 0.038 | 0.064 | 0.009 | 0.150 | 0.009 | 0.24 |
| NP3EO | 203 | 0 | 0.010 | 0.000 | 0.010 | 0.010 | 0.010 | 0.01 |
| OP | 7 | 0 | 0.010 | 0.000 | 0.010 | 0.010 | 0.010 | |
| OPEO | 7 | 100 | 0.033 | 0.016 | 0.031 | 0.051 | 0.018 | 0.01 |
| All APE | 1,069 | 33 | 0.043 | 0.089 | 0.010 | 0.210 | 0.002 | 0.05 |
| Mussels | n | Freq det (%) | Mean (μg/g wwt) | SD (μg/g wwt) | Median (μg/g wwt) | 95th percentile (μg/g wwt) | Min (μg/g wwt) | Max (μg/g wwt) |
| NP | 122 | 90 | 0.155 | 0.164 | 0.135 | 0.405 | 0.001 | 0.600 |
| NP1EO | 122 | 74 | 0.032 | 0.060 | 0.006 | 0.148 | 0.002 | 0.200 |
| NP2EO | 122 | 80 | 0.296 | 0.705 | 0.007 | 1.548 | 0.000 | 2.566 |
| NP3–16EO | 168 | 45 | 1.780 | 3.910 | 0.002 | 8.968 | 0.002 | 18.923 |
| NPEC | 12 | 0 | 0.002 | 0.000 | 0.002 | 0.002 | 0.002 | 0.002 |
| OP | 12 | 0 | 0.002 | 0.000 | 0.002 | 0.002 | 0.002 | 0.002 |
| OPEC | 12 | 0 | 0.002 | 0.000 | 0.002 | 0.002 | 0.002 | 0.002 |
| All APE | 570 | 65 | 0.877 | 2.752 | 0.002 | 4.157 | 0.000 | 18.923 |

In each case, the authors indicated that untreated or minimally treated wastewater was discharged into the receiving water. NP was detected in 36% of the surface water samples, with average concentrations of 0.25 ± 0.49 μg/L. OP, OPEO, and OPEC were detected in 25–60% of the samples, although the average concentrations were approximately 6–12 times lower than NP and NP>1EO. The latter findings are consistent with the fact that OP has significantly lower production volumes than NP, and the predominant use of OP does not result in its entry to the wastewater stream.

Figure 6a illustrates the frequency distribution for NP and NPEO concentrations in surface waters from the Great Lakes Basin. Concentrations of NP ranged from

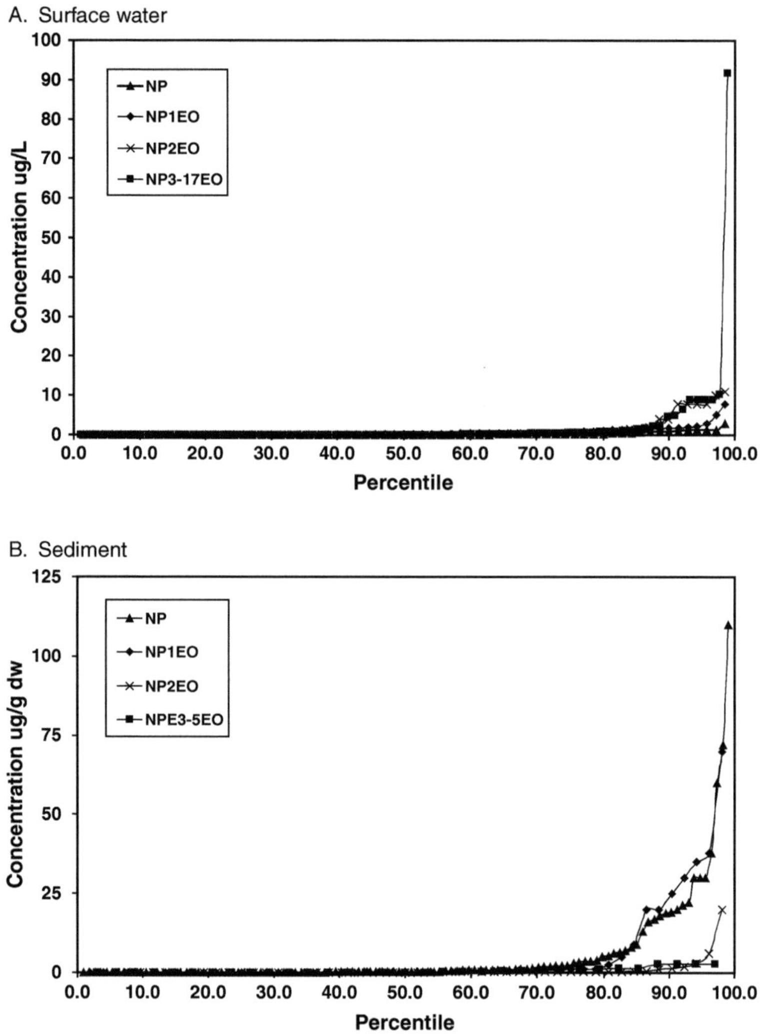


**Fig. 6** Frequency distribution of nonylphenol and nonylphenol ethoxylates in surface waters and sediments. **a** Surface water and **b** sediment

below the detection limit to as high as 3.0 μg/L (95th percentile = 1.3 μg/L). Of the NP concentrations reported, six exceeded 1 μg/L. The highest value (3.0 μg/L) was reported by Kolpin et al. (2002) for a sample taken from the Chicago Sanitary and Ship Canal. Kannan et al. (2003) reported one measured value of 1.1 μg/L downstream of the Kalamazoo (MI) wastewater treatment plant, with the remaining four values reported as non-detect. However, based on the detection limit reported for the method (<2.6 μg/L), an estimated concentration of 1.3 μg/L was entered in the database for use in the statistical analysis. Concentrations of low mole and high mole NPEO ranged from below the detection limit to maximum values of 7.8 and 91.7 μg/L, respectively. The 95th percentile concentrations for NP1EO, NP2EO, and NP3–17EO were 2.2, 7.8, and 8.9 μg/L, respectively. Of the 224 reported concentrations for low and high mole NPEO, 42 were greater than 1 μg/L. Nearly all of these were from samples taken downstream from wastewater treatment plant discharges. The highest NPEO concentration (91.7 μg/L) was reported by Mayer et al. (2007) in a sample obtained from Cootes Paradise, ON. As previously discussed, the authors attributed the high concentrations to a heavy rainfall event resulting in a large discharge from a combined sewer overflow on the day the samples were collected.

As summarized in Table 10b, concentrations of APEO and their metabolites have been reported for 166 sediment samples obtained from 1994 to 2002 from 34 locations throughout the Great Lakes Basin. Of the 361 values in the database, 71% were above the detection limit. NP and NPEO were frequently detected, with average concentrations ranging from 0.62 ± 0.94 to 5.6 ± 14.9 μg/g (dwt). OP and OPEO were also detected in the samples (71–100%), although the average concentrations were about ten fold lower.

Figure 6b illustrates the frequency distribution for sediment concentrations of NP and NPEO. NP concentrations ranged from below the detection limit to as high as 110 μg/g dwt (95th percentile = 30 μg/g dwt). Concentrations of low mole NPEO and high mole NPEO ranged from below the detection limit to maximum values from 2.9 to 70 μg/g dwt; the 95th percentile concentrations for NP1EO, NP2EO, and NP3–17EO were 32.5, 2.5, and 2.9 μg/g dwt, respectively. The highest concentrations of NP and NPEO were found in Hamilton Harbor and in the Detroit and Rouge Rivers downstream from wastewater treatment plant discharges, or in areas of heavy urbanization and industrial activity (Bennett and Metcalfe 1998, 2000; Kannan et al. 2001).

The concentrations of AP and APEO which have been reported in biological tissues are summarized in Table 10c. Much of the biological data have been reported as means or ranges of concentrations (min–max). Fish ($n = 284$) have been collected from both the Kalamazoo River (MI) and the Cuyahoga River (OH). Of the 1,069 values, 33% were above the detection limit. Concentrations of NP and the low mole ethoxylates (NP1–3EO) ranged from below the detection limit to maximum concentrations of 0.55 μg/g wwt. Concentrations of NP and NPEO have been reported in mussels collected from a reference lake near Montreal, QC, as well as for specimens submerged near wastewater treatment plant discharges in the St. Lawrence River, Hamilton Harbor, and the Detroit River. NP and low mole NPEO concentrations in mussels from the reference lake were below the detection limit, whereas detectable amounts of NP5–8EO (0.4–7.6 μg/g wwt) were reported. Concentrations of higher mole NPEO (NP>5EO), ranging from below the detection limit to 18.9 μg/g wwt, were reported in mussels submerged in the St. Lawrence river both upstream and downstream of the Montreal, QC, wastewater treatment plant. The absence of NP and low mole ethoxylates is consistent with the fact that the Montreal wastewater receives only primary physico-chemical treatment. Elsewhere, including mussels deposited in Hamilton Harbor and in the Detroit River downstream of the Windsor, ON, and Detroit, MI wastewater treatment plant discharges, maximum NP and NPEO concentrations were less than 1 μg/g wwt.

Regulatory criteria or guidelines have been developed in both the United States and Canada (Table 2). In the United States., the National Ambient Water Quality Criteria for NP in fresh water are 28 and 6.6 μg/L for acute and chronic exposures, respectively (US EPA 2005). In Canada, an Environmental Quality Guideline of 1 μg/L has been established for NP and NPEO in fresh water (Canadian Council of Ministers of the Environment (CCME) 2001). Considering the frequency distributions shown in Fig. 6a, all NP concentrations are below the US criterion, and 90% are below the Canadian guideline. In Canada, however, the guideline applies to aggregate NP equivalent concentrations, and the relative toxicity factors used to derive NP equivalent concentrations have been defined (CCME 2001). Lower mole NP$n$EO ($n = 1$–8) have been estimated to be less toxic than NP by a factor of two, and the higher mole NP$n$EO ($n > 9$), and NPEC were less toxic than NP by a factor of 200 (Servos et al. 2003). A similar approach was used for assigning relative toxicity factors for OPEO/OPEC, where OP was assigned a toxicity factor equal to NP. This methodology assumes that all compounds act with the same mode of action, and that the dose responses are additive. Based on the frequency distributions for AP and APEO in the Great Lakes, 78% of the NP equivalent concentrations in water are below the Canadian guideline. Although the US EPA has not specified a criterion based on aggregate concentrations, 95% of the NP equivalent concentrations are below 6.6 μg/L. These results are consistent with the recent work of Klečka et al. (2007), who reported the results of a statistical analysis for AP and APEO exposures in US aquatic environments.

Sediment quality guidelines have also been developed for Canada, but not in the United States. The current provisional Canadian guideline is 1.4 mg/kg dwt for NP equivalent concentrations (CCME 2001). Considering the frequency distributions, 69% of the NP equivalent sediment concentrations are below the guideline. A review of the database for the sites exceeding the guideline indicated that many of those with the highest concentrations are Canadian and are located in the immediate vicinity of wastewater treatment plant discharges. These sites (Windsor, Burlington, Hamilton, and Toronto) were sampled repeatedly during the period from 1995 to 1998 by several investigators (Bennett and Metcalfe 1998, 2000; Bennie et al. 1997). NP-equivalent sediment concentrations at these locations range from 16 to as high as 115 mg/kg dwt. Since the sediment samples were collected prior to implementation of the sediment quality guideline, it is unclear whether changes in regulations have had an impact on sediment exposures at these locations or not.

In summary, concentrations of AP and APEO have been reported for various media (water, sediment, biota) in samples collected throughout the Great Lakes Basin. Collectively, the frequencies with which the compounds have been detected range from 33 to 71%. Concentrations are generally low, with the exception of samples collected in the immediate vicinity of wastewater treatment plant discharges. Surface water concentrations were compared with United States and Canadian regulatory criteria and guidelines. None of the samples exceeded the US EPA Water Quality Criterion for NP. In contrast, 22% of the samples exceeded the NP-equivalent Canadian Water Quality Guideline. Sediment concentrations exceeded the NP-equivalent Canadian Sediment Guideline in 31% of the samples.

## 3.5 Synthetic Musks

Synthetic musks are used extensively in perfumes, cosmetics, detergents, cleaning products, and other personal care products. These compounds are commonly divided into two groups: nitro musks and polycyclic musks. There are other categories of synthetic fragrances, but they are used in much smaller quantities and therefore are not discussed here. The nitro musks include musk ketone (1-[4-(1,1-dimethyl-ethyl)-2,6-dimethyl-3,5-dinitrophenyl]-ethanone) and musk xylene (1-(1,1-dimethylethyl)-3,5-dimethyl-2,4,6-trinitrobenzene). The most common musks include 1,3,4,6,7,8-hexahydro-4,6,6,7,8,8-hexamethylcyclopenta(*g*)-2-benzopyran (HHCB), 6-acetyl-1,1,2,4,4,7-hexamethyltetralin (AHTN), 5-acetyl-3-isopropyl-1,1,2,6-tetramethylindane (ATII), 6-acetyl-1,1,2,3,3,5-hexamethylindane (AHMI), 4-acetyl-6-*tert*-butyl-1,1-dimethylindane (ADBI), and 6,7-dihydro-1,1,2,3,3-pentamethyl-4(5*H*)-indanone (DPMI) (Rimkus 1999). Nitro musks were first produced commercially in the 1900s as a substitute for the natural musk compounds, which were difficult to derive from their main source, the musk deer. In the 1950s, polycyclic musks were introduced commercially as another alternative to the natural musks. In 1996, worldwide production of synthetic musks was 8,000 t, with 5,600 t of this comprising polycyclic musks; of the latter, HHCB and AHTN accounted for 95% of the production (Gatermann et al. 2002).

Unlike their natural counterparts, synthetic musks have physical–chemical properties similar to those of other hydrophobic and semi-volatile organics, which are known to bioaccumulate and biomagnify in aquatic organisms. The environmental distribution and biological concentrations for musk ketone and musk xylene have been shown to correlate with their high octanol–water partition coefficients ($K_{\mathrm{OW}}$; Rimkus et al. 1997). HHCB and AHTN have slightly lower $K_{\mathrm{OW}}$ values than the nitro musks (Rimkus 1999). Environmental concentrations of the synthetic musks were first reported in 1981 for European surface waters (Yamagishi et al. 1981) and have since been reported for sediments, aquatic biota, as well as in human adipose tissue and breast milk (Rimkus 1999; Rimkus et al. 1994; Winkler et al. 1998). Current studies from Europe and Canada suggest that effluents from wastewater treatment plants are a major source of synthetic musks that enter the environment (Dsikowitzky et al. 2002; Ricking et al. 2003; Simonich et al. 2000).

Concentrations of synthetic musks in the Great Lakes Basin have been reported in five papers, in which their concentrations in surface water, sediment, and biota were characterized. Information on the concentrations of synthetic musks in air and precipitation (e.g., Peck and Hornbuckle 2006) was not included in the analysis.

Peck and Hornbuckle (2004) characterized the concentrations of a variety of synthetic musk fragrances in the water column of western Lake Michigan and in the vicinity of Milwaukee, WI (locations not illustrated). Water samples ($n = 13$) were collected during the period from 1999 to 2001. Results of the water analysis showed that musks are found primarily in dissolved phase, as opposed to adsorbed on suspended particles. Concentrations ranged from 0.029 for ADBI to 4.7 ng/L for HHCB. A mass balance analysis indicated that wastewater treatment plants are important sources, while volatilization plays a major role in the loss of synthetic musks from Lake Michigan.

Peck et al. (2006) examined the distribution of synthetic musks in single sediment cores taken in 2003 from Lakes Erie and Ontario (locations not illustrated). Analysis of surface sediments showed HHCB to be present at the highest concentrations in Lake Ontario (16 ng/g dwt), whereas ADBI was present in the lowest amounts (0.10 ng/g); musk xylene and AHMI were not detected in the samples. Analysis of samples taken from various depths showed that the concentrations of synthetic musk fragrances have been increasing with time. The results suggested that the rate of increase in deposition of these compounds in Great Lakes sediments corresponded with the rates at which production increased.

O'Toole and Metcalfe (2006) analyzed fish ($n = 22$) from the Great Lakes for the presence of a variety of synthetic musks. Various fish species were collected in 2002 from Hamilton Harbor in western Lake Ontario and from western Lake Erie and the Detroit River (locations not illustrated). Fish from Hamilton Harbor showed maximum total musk concentrations of $498.0 \pm 327.6$ ng/g wwt, with nearly 80% attributed to HHCB. Fish from the Detroit River and western Lake Erie had lower total musk concentrations (maximum of $20.1 \pm 4.0$ ng/g wwt), with HHCB again being the dominant compound detected. Of the synthetic musks detected in the various fish, the synthetic musks HHCB, AHTN, and ATII were most often detected. None of the fish had detectable concentrations of the nitro musks.

Concentrations of HHCB, AHTN, musk ketone, and musk xylene were recently reported for water samples collected in the spring of 2005 from Hamilton Harbor, ON, and from the North Sea (Andresen et al. 2007). Total musk concentrations ranged from 0.3 to 3 ng/L in samples ($n = 14$) taken from the North Sea, with higher concentrations found in nearshore areas. Considerably higher concentrations were detected in the samples ($n = 4$) from Hamilton Harbor. For example, maximum concentrations of HHCB and AHTN were 41 and 5.5 ng/L, respectively. At both locations, the concentrations declined with increasing distance from the source. The authors noted that some of the musk fragrances exhibit half-lives exceeding the residence times of the water body, and thus could be considered persistent in these ecosystems.

As previously discussed, Tertuliana et al. (2008) measured the concentration of a wide variety of organic contaminants in sediment samples collected in the Tinkers Creek watershed from locations above and downstream of wastewater discharges. AHTN and HHCB were detected in over 50% of the samples ($n = 18$) at maximum concentrations of 80 and 390 ng/g dwt, respectively.

A statistical analysis of synthetic musk concentrations in the Great Lakes watershed was performed using a total of 465 data points. Concentrations in water ($n = 375$) represented the majority of the data. Limited information was available for the concentrations in biota ($n = 42$) and sediment ($n = 48$). As summarized in Table 11a and b, the frequencies of detection ranged from 16 to 100% for most compounds in the media, except for DPMI, which was not detected in any of the samples. Detection frequencies for the total musks were lowest in water samples (37%), as would be predicted from their hydrophobic nature and volatility. The musks were detected most frequently in biota (68%), followed by sediment (50–58%; data not shown). Of the various synthetic musks, HHCB and AHTN were most frequently detected.

Concentrations in the water column were generally low; mean concentrations for total musks were $2.2 \pm 12.9$ ng/L (Table 11a). HHCB and AHTN were frequently detected in the water samples, with concentrations ranging from of 0.08 to 180 ng/L and from 0.08 to 28 ng/L, respectively. Musk ketone was also present at comparable levels (range from 0.03 to 57.3 ng/L). The other synthetic musks were detected at lower concentrations (<1 ng/L).

Calculated mean concentrations for total synthetic musks in biota were $105.6 \pm 122.2$ ng/g wwt, and ranged from 5.6 to 498 ng/g wwt (Table 11b). The highest single analyte detected in biota was HHCB (mean $= 80.3 \pm 96.0$ ng/g), representing about 76% of the total. The higher contributions for HHCB in biota are consistent with higher production volumes for this compound (Rimkus 1999). The mean concentration for three of the other synthetic musks, ATII, ADBI, and AHMI, were all 20–50-fold lower.

Limited data were available for sediment concentrations in the Great Lakes Basin. Based on the analysis of single sediment cores from both Lake Ontario and Lake Erie, concentrations of the various musks ranged from below the detection limit to 16.0 ng/g dwt. AHTN and HHCB were detected in sediments from Tinkers Creek at maximum concentrations of 80 and 390 ng/g dwt, respectively.

In general, the available data suggest that concentrations of the synthetic musks are somewhat higher in biota than in the other environmental matrices, consistent with their partitioning behavior. For all matrices, HHCB was the most frequently detected analyte as well as the one exhibiting the highest single analyte concentrations, reflecting the current status of the production volume of synthetic musks.

In the early 2000s, musk xylene was voluntarily removed from the market in some European countries after a report showed the compound causes hepatic toxicity in mice (Lehman-McKeeman et al. 1999). In 1999, the Centers for Disease Control supported research which showed developmental growth effects caused by musk xylene (Christian et al. 1999).

**Table 11** Summary statistics for synthetic musks

| | *n* | Freq det (%) | Mean (ng/L) | SD (ng/L) | Median (ng/L) | 95th percentile (ng/L) | Min (ng/L) | Max (ng/L) |
|---|---|---|---|---|---|---|---|---|
| *A. Water* | | | | | | | | |
| AHTN | 50 | 66 | 1.525 | 4.800 | 0.334 | 4.770 | 0.075 | 28.000 |
| HHCB | 50 | 70 | 8.440 | 31.210 | 0.817 | 27.585 | 0.075 | 180.000 |
| Acetyl-isopropyl-tetramethylindane (ATII) | 44 | 30 | 0.078 | 0.049 | 0.075 | 0.113 | 0.018 | 0.361 |
| Acetyl-*tert*-butyl-dimethylindane (ADBI) | 44 | 16 | 0.071 | 0.021 | 0.075 | 0.075 | 0.018 | 0.155 |
| Acetyl-hexamethylindane (AHMI) | 44 | 41 | 0.132 | 0.117 | 0.075 | 0.292 | 0.045 | 0.617 |
| Dihydro-pentamethyl-indanone (DPMI) | 43 | 0 | 0.075 | 0.000 | 0.075 | 0.075 | 0.075 | 0.075 |
| Musk ketone | 50 | 50 | 5.990 | 14.083 | 0.075 | 42.171 | 0.030 | 57.335 |
| Musk xylene | 50 | 64 | 0.061 | 0.048 | 0.075 | 0.100 | 0.006 | 0.260 |
| Total musks | 375 | 37 | 2.177 | 12.904 | 0.075 | 37.066 | 0.006 | 180.000 |
| | *n* | Freq det (%) | Mean (ng/g wwt) | SD (ng/g wwt) | Median (ng/g wwt) | 95th percentile (ng/g wwt) | Min (ng/g wwt) | Max (ng/g wwt) |
| *B. Biota* | | | | | | | | |
| ATHN | 8 | 100 | 19.16 | 26.26 | 5.40 | 70.73 | 1.50 | 96.80 |
| HHCB | 8 | 100 | 80.34 | 95.85 | 8.85 | 303.71 | 1.70 | 391.10 |
| ATII | 8 | 63 | 3.70 | 4.40 | 0.45 | 9.62 | 0.20 | 9.90 |
| ADBI | 8 | 25 | 1.50 | 1.70 | 0.10 | 5.94 | 0.10 | 6.50 |
| AHMI | 8 | 50 | 1.45 | 4.42 | 0.15 | 5.42 | 0.10 | 5.80 |
| Total musks | 40 | 68 | 105.58 | 122.25 | 14.80 | 403.26 | 5.60 | 498.00 |

Risk assessments for musk xylene (EUC 2005a), musk ketone (EUC 2005b), AHTN (EUC 2008b), and HHCB (EUC 2008c) have been recently completed in Europe. PNECs have been developed for use in the calculation of risk quotients (Table 2). A comparison of the reported maximum concentrations of musk xylene, musk ketone, AHTN, and HHCB musk xylene, in environmental media from the Great Lakes, indicates that all values are below the PNEC values.

### *3.6 Perfluorinated Surfactants*

The presence of perfluorinated surfactants in the environment and in biological tissues (Giesy and Kannan 2001; Giesy et al. 2010; Kannan et al. 2002) can be attributed to their widespread use over the past 50 years in a broad range of applications, including use as surface treatments for carpets, fabrics, and leathers for stain resistance, as well as coatings on various paper products. These surfactants have also found applications in semiconductor manufacturing, metal plating, and are used as components of hydraulic fluids in aviation, and in firefighting foams. There is growing concern over the environmental presence of perfluorinated surfactants, because representatives like perfluorooctane sulfonate (PFOS) have been shown to resist hydrolysis, photolysis, and biodegradation under environmental conditions. Perfluorinated surfactants, including PFOS and perfluorooctanoic acid (PFOA), are also of concern because they have been found to accumulate in biological tissues (Kannan et al. 2002). PFOS and related fluoroalkyl substances have been detected in blood plasma of non-occupationally exposed humans collected from around the world and have even been found in sparsely populated regions that have no apparent sources (Schultz et al. 2003).

Perfluorinated surfactants are currently synthesized by two distinct processes, namely electrochemical fluorination (ECF) and telomerisation, which yield different types of fluorinated reaction products (Hekster et al. 2002). Electrochemical fluorination products are branched or straight-chain polymers that contain a sulfonyl group; perfluorooctanesulfonyl fluoride (POSF) is the most important manufacturing intermediate. POSF is a precursor of many fluorinated sulfonyl surfactants found in the environment, including PFOS, perfluorooctane sulfinate (PFOSulfinate), perfluorooctane sulfonamide (FOSA or PFOSA), perfluorooctane sulfonamido acetic acid (PFOSAA), *N*-ethylperfluorooctane sulfonamido acetic acid (*N*-EtFOSAA), and *N*-ethylperfluorooctane sulfonamidoethanol (*N*-EtFOSE). The telomerisation process is distinct from electrochemical fluorination as it produces only linear polymer chains that are typically fluorotelomer olefins, epoxides, or fluorotelomer alcohols (FTOHs). The alcohols are precursors to many perfluorinated carboxylic acids (PFAs), such as PFOA and perfluorononanoic acid (PFNA). In commercial products, the fluoroalkyl chain length can vary from 4 to 20 carbon atoms, although most products contain chain lengths between 6 and 10. The homologous series of PFAs found in the environment include C4 (perfluorobutyric acid; PBFA), C5 (perfluoropentanoic acid; PFPeA), C6 (perfluorohexanoic acid, PFHxA), up to and including C15 (perfluoropentadecanoic acid, PFPA). Short-chain fluorinated acids such as trifluoroacetic acid (TFA) and the C3–C9 PFAs are also reported to be atmospheric oxidation products of fluorotelomer alcohols and perfluorosulfonamido alcohols (Martin et al. 2006).

The complex structure of perfluorinated surfactants makes it difficult to predict their environmental behavior solely from examining their physical–chemical properties. These structures have both a hydrophobic head and a hydrophilic carbon chain tail. Although this structural configuration lends itself to being a highly effective surfactant, it also thwarts scientific attempts to understand the potential environmental fate and bioconcentration potential for members of the class. For example, unlike organic chemicals that bioaccumulate by partitioning into lipids, perfluorinated surfactants have been shown to bind to proteins and are slowly excreted (Kannan et al. 2005a). Because PFOS has the potential to accumulate in the liver, it has been shown to be a potential carcinogen in rats and has displayed significant increases in hepatocellular adenomas in animals exposed orally to dietary concentrations of 20 mg/kg (Organisation for Economic Co-operation and Development (OECD) 2002).

The present analysis focused on those studies which reported concentrations in surface waters and biota. However, a number of studies were identified in the literature search which included analysis of perfluorinated surfactants in air and precipitation (e.g., Boulanger et al. 2005; Scott et al. 2006).

Boulanger et al. (2004) were among the first to analyze for perfluorinated surfactants in the waters of the Great Lakes. Surface water samples were collected in August 2003 from four locations in Lake Erie ($n = 8$) and four locations in Lake Ontario ($n = 8$) and were analyzed for a PFOS, PFOA, and six other PFOS precursors. Concentrations of PFOS and PFOA in Lake Erie and Lake Ontario ranged from 11 to 47 ng/L and from 15 to 121 ng/L. The analysis also revealed the presence of various PFOS precursors, including *N*-EtFOSAA (4.2–11 ng/L), and FOSA (0.6–1.3 ng/L). PFOSAA and PFOSulfinate were detected in some samples at concentrations ranging from below the detection limit to maximum levels of 6.5 and 17 ng/L, respectively. *N*-EtFOSE and *N*-EtFOSAA were not detected in surface waters from the two lakes.

In a related study, Boulanger et al. (2005) used the data from their previous work to develop a mass budget for the eight perfluorooctane surfactants found in Lake Ontario. The mass balance showed that inflows from Lake Erie, as well as contributions from wastewater treatment plants, were the major sources of perfluorooctane surfactants in Lake Ontario. Outflow through the St. Lawrence River was shown as the predominant loss mechanism. Based on the analysis, the authors concluded that the steady state and measured mean concentrations in the lake are the same at the 95% confidence level.

Scott et al. (2006) described the development of a sensitive analytical method for PFAs and application of these methods to analyze surface water samples collected from varying locations in the Great Lakes Basin. Lake water samples were obtained in 2001 at varying depths from Lakes Superior (4 and 260 m; $n = 6$), Huron (125 m; $n = 2$), and Ontario (4–200 m; $n = 9$). The total PFA concentrations found in Lake Ontario were relatively constant at depths from 4 to 150 m, ranging from 112 to 162 ng/L, whereas total concentrations in the deeper water samples were higher (233 ng/L). TFA was the dominant PFA detected; TFA concentrations ranged from 100 to 200 ng/L, with the levels increasing with depth. Concentrations of the other analytes (C3–C8; PFPrA to PFOA) were lower, ranging from below the detection limit to 8.9 ng/L. PFOA and PFNA concentrations in Lakes Superior and Huron ranged from below the detection limit to 1.8 ng/L.

Martin et al. (2004) analyzed the bioaccumulation and trophic magnification of a variety of perfluorinated compounds in a Lake Ontario food web. Biological specimens consisting of invertebrates (*Diporeia* and *Mysis* spp.), forage fish (slimy sculpin, alewife, rainbow smelt), and lake trout were collected in 2002 and analyzed for PFOS and its precursor FOSA, as well as the homologous series of PFAs (C8–C15). Total PFOS concentrations ranged from 50 to 460 ng/g wwt and were highest in *Diporeia* and sculpin. The distribution in the food web was similar for the PFAs. Total PFA concentrations ranged from 12 to 260 ng/g wwt and were also highest in *Diporeia* and sculpin. Total PFOS and PFA concentrations in the trout were 186 and 30 ng/g wwt, respectively. Analysis of concentration versus trophic level for the various analytes did not reveal any significant associations when all of the organisms were included in the regression analysis. For the PFAs, the benthic macroinvertebrate (*Diporeia*) and sculpin showed wide and consistent divergence from the other organisms. When these organisms were excluded, strong associations with trophic level were observed. Trophic magnification factors exceeded unity for PFOS, perfluorodecanoic acid (PFDA), perfluoroundecanoic acid (PFUnA), and perfluorotridecanoic acid (PFTrA), ranging from 2.45 to 5.88. Trophic magnification factors were equal to or less than one for FOSA, PFOA, PFNA, and PFDoA. Archived lake trout samples from the Great Lakes Specimen Bank were also analyzed for PFOS to determine any temporal trends in Lake Ontario. PFOS concentrations increased from 43 to 180 ng/g during the period from 1980 to 2001. The authors noted that the increase was not linear and may have been influenced by the invasion and proliferation of zebra mussels.

Kannan et al. (2005b) examined the trophic transfer of PFOS, PFSOA, perfluorohexane sulfonate (PFHxS), and PFOA in a wide range of aquatic and terrestrial wildlife collected from Michigan and the surrounding region in 1998–2001. Average PFOS concentrations in amphipods (1.6 ng/g wwt), zebra mussels (1.4 ng/g wwt), and crayfish (3.5 ng/g wwt) were approximately 1000-fold greater than those in the surrounding water (1.9–3.9 ng/L). PFOS concentrations in round gobies (mean = 10.6 ng/g wwt) and smallmouth bass (mean = 9.1 ng/g wwt) were 2- to 4-fold greater than those measured at lower trophic levels. Similar trends were noted for PFOSA at the lower trophic levels. PFOS accumulation in predatory fish was determined on the basis of liver analysis, since the compound has been shown to preferentially accumulate in this tissue. Livers from lake whitefish and Chinook salmon contained average PFOS concentrations of 67 and 100 ng/g wwt, respectively, which is approximately 6–10-fold greater than that detected in round gobies, and 10–20-fold higher than present in zebra mussels and amphipods. Carp from the Saginaw Bay contained PFOS concentrations as high as 297 ng/g wwt in muscle tissue. Although the concentrations were quite variable, concentrations in mink and bald eagles were, on average, five to tenfold greater than those measured in salmon, carp, or snapping turtles. Overall, the results suggest a bioaccumulation factor of approximately 1,000 for benthic invertebrates, and a biomagnification factor of 10–20 for mink and bald eagles. Eggs from several fish collected in the study contained detectable concentrations of PFOS (64 ng/g wwt), suggesting the potential for oviparous transfer. In contrast, although PFOA concentrations in water ranged from 4.4 to 14.7 ng/L, the compound was detected in few of the biological specimens, suggesting a biomagnification potential lower than PFOS.

Sinclair et al. (2006) reported the occurrence of PFOS and several related compounds in water, fish, and birds from New York State. Water samples ($n = 51$) were obtained from nine water bodies located throughout the state in 2004. In addition, sport fish ($n = 66$) were collected from 20 inland lakes, and 10 species of waterfowl ($n = 87$) were obtained from the Niagara River region. Residues of PFOS, PFOA, and PFHxS were ubiquitous in the surface waters. Average PFOA concentrations ranged from 14 to 49 ng/L and were typically higher than concentrations for PFOS (means range from 1.6 to 6.4 ng/L) or PFHxS (means range from 0.9 to 7.4 ng/L). Elevated concentrations of PFOS were found in Lake Onondaga (198–1,090 ng/L), and higher concentrations of PFOA (22–173 ng/L) were detected in the Hudson River. Of the various analytes, PFOS was the most abundant perfluorinated compound present in all of the wildlife samples. PFOS concentrations in fish and bird livers ranged from 9 to 315 ng/g wwt and from 11 to 882 ng/g wwt, respectively, and the concentrations were greater in piscivorous birds. PFOA and PFOSA were also detected in 62–90% of the fish livers at concentrations ranging from below the detection limit to 11.4 ng/g wwt. However, PFOA and PFOSA residues were not found in any of the bird samples. Based on the results of the study, a bioconcentration factor of 8,850 was estimated for PFOS accumulation in fish, and a biomagnification factor of 8.9 was estimated for the accumulation of PFOS in the common merganser.

The spatial distribution of various perfluoroalkyl contaminants in lake trout from the Great Lakes was recently reported by Furdui et al. (2007). Lake trout of the same age class (4 years) were collected in 2001 from Lakes Superior, Michigan, Huron, Erie, and Ontario. Between 6 and 10 fish were obtained from each lake, and whole fish homogenates were analyzed for a broad range of perfluorinated sulfonates (PFHxS, PFOS, PFDS, and PFOSA) and carboxylates (perfluoroheptanoic acid (PFHpA) to perfluoropentadecanoic acid (PFPA); C7–C15). Fish from Lake Superior contained the lowest total concentrations ($13.1 \pm 1$ ng/g wwt), while the highest levels were found in samples from Lake Erie ($152 \pm 14$ ng/g wwt). Fish from Lakes Ontario and Huron showed similar total concentrations, $60 \pm 5$ ng/g wwt and $58 \pm 10$ ng/g wwt, respectively, whereas samples from Lake Michigan were lower ($27 \pm 3$ ng/g wwt). The predominant perfluorinated compounds detected were PFOS followed by perfluorodecane sulfonate (PFDS); mean concentrations for these analytes ranged from 5 to 121 ng/g wwt and from 0.7 to 9.8 ng/g wwt, respectively. Average concentrations for the perfluorinated carboxylic acids were lower, ranging from below the detection limit to 4.9 ng/g wwt. Of the PFAs detected, PFOA and PFDA were typically present at higher concentrations.

The statistical analysis of perfluorinated surfactant concentrations in the Great Lakes watershed was based on the data obtained from seven publications in which the concentrations in surface water and biological tissues were reported. The samples were collected during the period from 1998 to 2003, and the locations are illustrated in Fig. 7a and b. Water samples were collected from Lakes Erie, Huron, Ontario, and Superior. Biological samples have been obtained from locations throughout the Great Lakes watershed. For the analysis, the various perfluorinated compounds were subdivided into two major categories based on the manufacturing process used for their synthesis.

As shown in Table 12a and b, detection frequencies for total perfluorinated sulfonyls (PFS) were highest in water (65%), followed by biota (47%). Detection frequencies for the PFAs were higher in all media in which they were analyzed.

PFS and PFA were prevalent in surface waters, with all but four compounds detected in 50% or more of the samples. The total concentrations of PFS and PFA in water were low, with mean concentrations of 0.0126 ± 0.0662 μg/L and 0.0206 ± 0.0377 μg/L, respectively (Table 12a). PFOS and PFOA were the predominant analytes detected in water. Analysis of the database for information on the spatial distribution of perfluorinated surfactants in Great Lake waters suggests that the highest concentrations are found in Lake Ontario.

PFS and PFA were widely detected in a variety of aquatic and terrestrial wildlife, as shown in Table 12b. The frequencies with which the various analytes were detected ranged from 4 to 100%. The total PFS concentrations in biota were higher than those of PFA; their mean concentrations ranged from 119.2 ± 1,170.9 ng/g wwt to 7.19 ± 12.44 ng/g wwt, respectively. PFOS and PFOA were the predominant contaminants detected in biota, with concentrations ranging from below the detection limit to maximum values of 18,000 and 90 ng/g wwt, respectively. From the analysis of lake trout collected from each of the five Great Lakes, the concentrations of PFS and PFA are highest in Lake Erie, followed by Lakes Ontario and Huron, with Lake Michigan and Lake Superior showing lower concentrations. From the analysis of archived fish, temporal trends in PFOS concentrations in biota appear to be increasing with time (Martin et al. 2004).

Risk assessments for PFOS and PFOA have been conducted by authorities in various geographies (Department for Environment, Food and Rural Affairs; DEFRA 2004; EC 2006a; Hekster et al. 2002). Environment Canada has derived an estimated no effect value (ENEV) of 0.491 μg/L for PFOS in freshwater. PNECs have also been derived in Europe for PFOS in fresh water (2.5 μg/L) and for secondary poisoning by ingestion of food (0.0167 mg/kg food). The PNEC for water is comparable to the proposed Dutch Quality Objective for water of 3.8 μg/L. Comparison of the available data for PFOS and PFOA concentrations in water versus the various regulatory standards (Table 2) suggests that all exposures are well below the no effect values derived in Europe. Mean and 95th percentile values for water are also below the Canadian ENEV, although the maximum reported PFOS concentration (0.76 μg/L) is higher. It is important to note that the highest concentration in the database was measured in Lake Onondaga in New York, which, according to the authors, is a Superfund site that is influenced by several industries that exist along

A. Surface waters

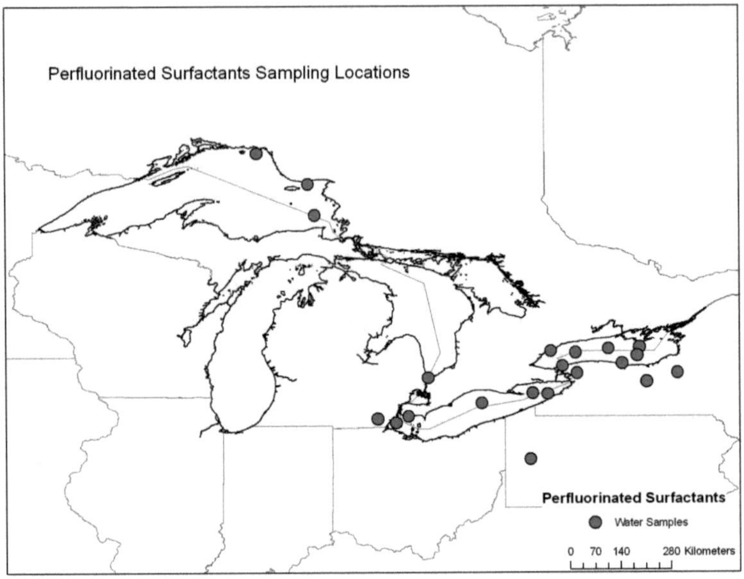


B. Biota

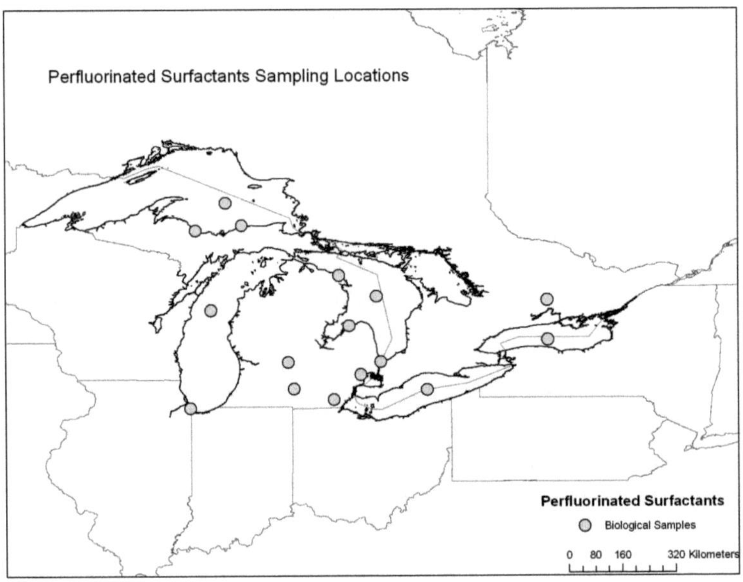


**Fig. 7** Sampling locations for perfluorinated surfactants. **a** Surface waters and **b** biota

**Table 12** Summary statistics for perfluorinated surfactants

| | *n* | Freq det (%) | Mean (μg/L) | SD (μg/L) | Median (μg/L) | 95th percentile (μg/L) | Min (μg/L) | Max (μg/L) |
|---|---|---|---|---|---|---|---|---|
| *A. Water* | | | | | | | | |
| Perfluorohexane sulfonic acid (PFHxS) | 11 | 82 | 0.0017 | 0.0020 | 0.0012 | 0.0050 | 0.0005 | 0.0074 |
| Perfluorooctane sulfonic acid (PFOS) | 27 | 100 | 0.0550 | 0.1428 | 0.0310 | 0.1039 | 0.0016 | 0.7560 |
| Perfluorooctane sulfinate (PFOSulfinate) | 16 | 69 | 0.0031 | 0.0057 | 0.0011 | 0.0173 | 0.0001 | 0.0180 |
| Perfluorooctane sulfonamide (PFOSA) | 18 | 78 | 0.0014 | 0.0014 | 0.0011 | 0.0050 | 0.0001 | 0.0050 |
| Perfluorooctane sulfonamido acetic acid (PFOSAA) | 16 | 38 | 0.0006 | 0.0016 | 0.0001 | 0.0020 | 0.0001 | 0.0065 |
| *N*-Ethylperfluorooctane sulfonamide (*N*-EtFOSA) | 16 | 6 | 0.0003 | 0.0001 | 0.0003 | 0.0003 | 0.0001 | 0.0003 |
| *N*-Ethylperfluorooctane sulfonamido acetic acid (*N*-EtFOSAA) | 16 | 94 | 0.0066 | 0.0036 | 0.0073 | 0.0110 | 0.00001 | 0.0110 |
| *N*-Ethylperfluorooctane sulfonamido ethanol (*N*-EtFOSE) | 16 | 6 | 0.0010 | 0.0003 | 0.0011 | 0.0011 | 0.0001 | 0.0011 |
| Total perfluorinated sulfonyls (PFS) | 136 | 65 | 0.0126 | 0.0662 | 0.0011 | 0.0418 | 0.00001 | 0.7560 |
| Trifluoroacetic acid (TFA) | 8 | 100 | 0.1344 | 0.0371 | 0.1200 | 0.1913 | 0.1000 | 0.2000 |
| Perfluoropropionic acid (PFPrA) | 9 | 100 | 0.0036 | 0.0020 | 0.0031 | 0.0068 | 0.0023 | 0.0089 |
| Perfluorobutyric acid (PFBA) | 9 | 100 | 0.0054 | 0.0014 | 0.0049 | 0.0077 | 0.0045 | 0.0091 |
| Perfluoropentanoic acid (PFPeA) | 9 | 100 | 0.0017 | 0.0019 | 0.0013 | 0.0047 | 0.0004 | 0.0065 |
| Perfluorohexanoic acid (PFHxA) | 9 | 100 | 0.0017 | 0.0009 | 0.0012 | 0.0032 | 0.0011 | 0.0037 |
| Perfluoroheptanoic acid (PFHpA) | 9 | 78 | 0.0015 | 0.0015 | 0.0008 | 0.0035 | 0.0001 | 0.0036 |
| Perfluorooctanoic acid (PFOA) | 44 | 93 | 0.0218 | 0.0195 | 0.0190 | 0.0499 | 0.00013 | 0.0700 |
| Perfluorononanoic acid (PFNA) | 8 | 38 | 0.0002 | 0.0001 | 0.0003 | 0.0003 | 0.00003 | 0.0004 |
| Total perfluorinated carboxylic acids (PFA) | 105 | 90 | 0.0206 | 0.0377 | 0.0035 | 0.1080 | 0.00003 | 0.2000 |

**Table 12** (continued)

| | *n* | Freq det (%) | Mean (ng/g wwt) | SD (ng/g wwt) | Median (ng/g wwt) | 95th percentile (ng/g wwt) | Min (ng/g wwt) | Max (ng/g wwt) |
|---|---|---|---|---|---|---|---|---|
| *B. Biota*[a] | | | | | | | | |
| PFHxS (PFHS) | 72 | 4 | 4.5 | 5.7 | 1.0 | 17.0 | 0.5 | 21.0 |
| PFOS | 82 | 85 | 328.2 | 1,998.6 | 13.5 | 449.8 | 1.0 | 18,000.0 |
| Perfluorodecane sulfonic acid (PFDS) | 5 | 100 | 3.3 | 3.8 | 1.6 | 8.5 | 0.7 | 9.8 |
| PFOSA (FOSA) | 82 | 43 | 18.1 | 32.7 | 2.6 | 70.3 | 0.3 | 180.0 |
| Total PFS | 241 | 47 | 119.2 | 1,170.9 | 4.1 | 168.0 | 0.3 | 18,000.0 |
| PFOA | 82 | 16 | 8.66 | 13.87 | 2.25 | 36.00 | 0.01 | 90.00 |
| PFNA | 11 | 100 | 10.26 | 7.29 | 2.80 | 45.00 | 0.57 | 57.00 |
| Perfluorodecanoic acid (PFDA) | 11 | 100 | 7.84 | 3.16 | 2.20 | 30.50 | 0.72 | 32.00 |
| Perfluoroundecanoic acid (PFUnA) | 11 | 100 | 9.76 | 3.77 | 2.70 | 40.00 | 0.74 | 41.00 |
| Perfluorododecanoic acid (PFDoA) | 11 | 100 | 3.91 | 1.46 | 1.80 | 14.00 | 0.37 | 14.00 |
| Perfluorotridecanoic acid (PFTrA) | 11 | 100 | 4.57 | 1.27 | 3.30 | 14.00 | 1.10 | 15.00 |
| Perfluorotetradecanoic acid (PFTeA) | 11 | 82 | 1.74 | 0.57 | 1.00 | 5.65 | 0.25 | 7.30 |
| Perfluoropentadecanoic acid (PFPA) | 11 | 64 | 0.67 | 0.46 | 0.46 | 1.54 | 0.25 | 1.90 |
| Total PFA | 159 | 53 | 7.19 | 12.44 | 1.60 | 36.00 | 0.01 | 90.00 |

[a] The results represent concentrations reported for multiple trophic levels (invertebrates, small fish, large fish, birds, mammals)

the lake (Sinclair et al. 2006). As a result, concentrations measured at this location may have no relevance to the Great Lakes. However, PFOS and total PFS concentrations in biota exceed the criteria for secondary poisoning from the ingestion of food. As summarized in Table 12b, mean and 95th percentile concentrations for PFOS and total PFS are 0.33 and 0.45 mg/kg wwt and 0.12 and 0.17 mg/kg wwt, respectively.

## 3.7 Polybrominated Diphenyl Ethers

PBDEs are a class of additive halogenated flame retardants used in numerous commercial products to impart fire retardant properties. The flame retardant within the polymer is not covalently bonded (to the polymer). Rather, these additive flame retardants are mixed with the polymer resin and therefore have the potential to continually migrate out of the final product. This tendency for PBDEs to leach out of treated materials or to enter the environment during production, formulation, and use has contributed to the detection of PBDE globally, as environmental contaminations. These compounds have become chemicals of emerging concern in recent years, because of their high production volume during the period from the 1970s to the 1990s and their detection in surface waters, sediments, air, biota, and humans. These compounds are hydrophobic, lipophilic, and therefore have the potential for accumulation in sediments and wildlife, with biomagnification occurring within the food web.

The most common PBDEs identified in the environment are comprised predominately of the tetrabrominated congener, BDE-47, followed by lower levels of two pentabrominated congeners, BDE-99 and BDE-100 (Rice et al. 2002). PBDEs are structurally similar to the PCBs, and thus the numbering of PBDE congeners follows the system used to identify PCB congeners. As a result of the similarities between PBDEs and PCBs, researchers have examined and compared concentrations in fish and sediment samples in attempts to correlate contaminant distributions and loading rates for these two compounds in the Great Lakes. Production of the commercial penta-PBDE and octa-PBDE products was banned in Europe in 2003 (EUC 2003c), and current efforts are focused on phasing out these brominated flame retardants in North America. With the regulatory ban of the penta-PBDE and octa-PBDE products, the production of other brominated flame retardants, such as decabromodiphenyl ether (deca-PBDE product), has increased. As summarized below, numerous researchers have attempted to track changes in PBDE concentrations and congener profiles. In many of these studies, attempts have been made to establish the temporal trends for these compounds in the Great Lakes, by using archived fish and egg samples that were collected as part of the Great Lakes Fish Monitoring Program or by the US Geological Survey Great Lakes Science Center. More recent studies have focused on the analysis of fresh fish and bird eggs along with sediment samples. The locations of the sampling sites are shown in Fig. 8a and b.

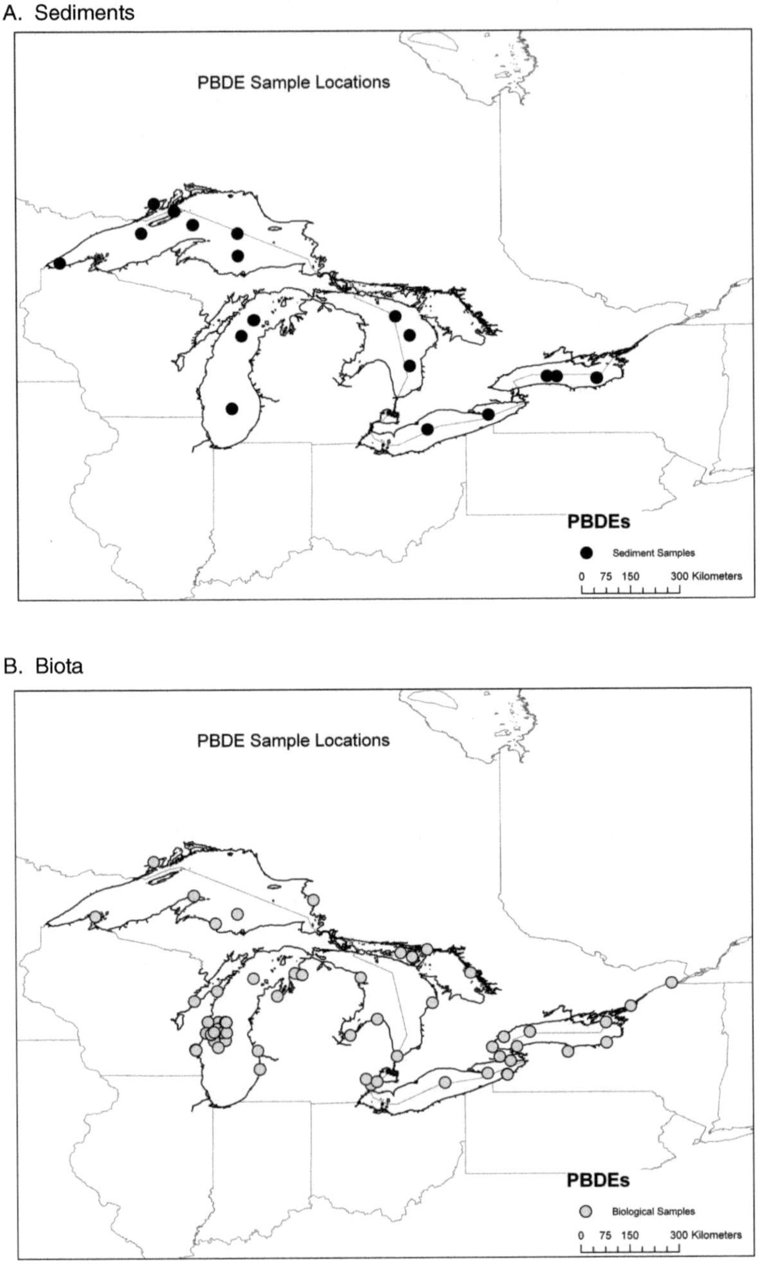


**Fig. 8** Sampling locations for polybrominated diphenyl ethers. **a** Sediments and **b** biota

The present analysis focused on those studies which reported concentrations in sediments and biota. However, a number of studies were identified in the literature search which described the analysis of brominated diphenyl ethers in the atmosphere over the Great Lakes (e.g., Gouin et al. 2005; Hoh and Hites 2005; Shen et al. 2006; Strandberg et al. 2001; Venier and Hites 2008).

Manchester-Neesvig et al. (2001) compared the levels of PCBs and PBDEs in coho (*Oncorhynchus kisutch*) and chinook (*Oncorhynchus tshawytscha*) salmon collected from two southern Lake Michigan tributaries in the fall of 1996. The average concentration for total PBDEs in salmon was 80.1 ng/g wwt and consisted predominantly of BDE-47, BDE-66, BDE-100, BDE-99, BDE-154, and BDE-153. This concentration was compared to the average total PCB concentration which was approximately 18 times higher at 1,450 ng/g wwt. When the most abundant single congeners for PBDEs and PCBs were compared, the values became more similar with an average BDE-47 concentration of 52.1 ng/g wwt, compared to 149 ng/g wwt for PCB-153. Since the concentrations of PBDE and PCB were correlated in individual fish, the authors suggested that PBDE concentrations are as prevalent as PCBs in Lake Michigan.

Luross et al. (2002) determined the concentration of two types of brominated flame retardants, including PBDEs and polybrominated biphenyls (PBBs) in a single age class of lake trout (*Salvelinus namaycush*), collected in 1997 from Lakes Superior, Huron, Erie, and Ontario. Total PBDE concentrations in lake trout were reported to be significantly higher in trout collected from Lake Ontario (95 ng/g wwt), than in all other lakes except Lake Superior (56 ng/g wwt). Concentrations for Lake Huron and Lake Erie were 50 and 27 ng/g wwt, respectively. The three most dominant congeners representing the penta-PBDE formulation, BDE-47, BDE-99, and BDE-100, accounted, respectively, for 57, 15, and 8% of the total PBDEs measured. Lake Huron trout had the highest levels of PBBs with maximum concentrations of 3.1 ng/g wwt. The authors concluded that higher concentrations of PBDEs versus PBBs may be attributed to a North American ban on PBB production that was implemented approximately 20 years ago. Differences observed in concentrations of these two flame retardants within the Great Lakes were attributed to probable different sources of input.

The historical and geographical distribution of 15 PBDE congeners and 1 PBB (PBB-153) was examined in archived lake trout (*S. namaycush*) and walleye (*Stizostedion vitreum*) collected from the Great Lakes during the sampling period 1980–2000 by Zhu and Hites (2004). Lake Michigan and Lake Ontario trout had the highest total PBDE concentrations ($\Sigma$PBDEs = BDE-47, BDE-99, BDE-100, BDE-153, BDE-154). An exponential increase in PBDE concentrations was shown from 1980 to 2000 in fishes from all five of the Great Lakes; this increase had a calculated doubling time of 3–4 years, which is similar to results reported by other studies (Norstrom et al. 2002). The authors reported no change in PBB-153 concentrations over the period of 1980–2000, except in Lake Huron where the concentrations have decreased.

Carlson and Swackhamer (2006) analyzed composite samples of 820 lake trout (*S. namaycush*), walleye (*S. vitreum*), steelhead (variant of rainbow trout,

*Oncorhynchus mykiss*), chinook (*O. tshawytscha*) and coho (*O. kisutch*) from the Great Lakes collected in 1999 and 2000, as part of the Great Lakes Fish Monitoring Program. Concentrations of PBDE congeners (355 ng/g wwt) were greatest in lake trout from Lake Michigan (sum of BDE-47, BDE-66, BDE-99, BDE-100, BDE-153, and BDE-154), followed by Lake Ontario and Lake Superior. The trout from Lake Huron had the lowest concentrations of these compounds. The authors reported on the differences in contamination patterns among lakes, between sites within a lake, and between fish from the same site.

Chernyak et al. (2005) evaluated time trends (1983–1999) for a variety of organochlorines and PBDEs, using the rainbow smelt (*Osmerus mordax*) as an indicator organism for detection of the biomagnification of these substances. The smelt collected in Lakes Michigan, Huron, and Superior between 1983 and 1999 were archived. The most abundant congener was BDE-47, and time trends for total PBDE (sum of BDE-47, BDE-99, BDE-100, and BDE-153) concentrations indicated exponential increases at all sites, with doubling times varying from 1.58 to 2.94 years in Lake Huron, 2.16 years for Lake Superior, and 1.75 years for Lake Michigan. The authors concluded that the initiation of build-up of PBDEs occurred in 1980.

Batterman et al. (2007) analyzed trends in PBDE concentrations and congener profiles in rainbow smelt (*O. mordax*), and lake trout (*S. namaycush*) collected in Lakes Michigan, Superior, Huron, Ontario, and walleye (*S. vitreum*) from Lake Erie between 1979 and 2005. Their analyses focused on the four most common congeners (BDE-47, BDE-99, BDE-100, and BDE-153). This research is an update to trend data determined by a previous analysis by Chernyak et al. (2005). Trend analysis indicated increasing concentrations of PBDE in fish from all Great Lakes starting in the early to mid-1980s. Although the doubling times were 2–4 years, differences were reported by congener, fish species, and source lake. From the more recent data, the authors concluded that PBDE concentrations are no longer exponentially increasing in the Great Lakes as was reported by other studies. The data suggest that accumulation rates are slowing and concentrations of penta and hexa-congeners may be decreasing. This trend was most evident in Lake Ontario and Lake Michigan, beginning in the mid-1990s.

Streets et al. (2006) examined the partitioning and bioaccumulation of PBDEs and PCBs in the Great Lakes and were the first to determine bioaccumulation factors (BAFs) in fish. The researchers analyzed 6 PBDE and 110 PCB congeners in lake trout collected as part of the Great Lakes Fish Monitoring Program during 2000, 2001, and 2002 from all five Great Lakes and water samples collected from five locations in Lake Michigan. Four PBDE congeners (BDE-47, BDE-66, BDE-99, and BDE-100) were detected in the dissolved phase of water samples and had concentrations ranging from 0.13 to 10 pg/L. In the particulate phase, three PBDE congeners (BDE-47, BDE-99, and BDE-100) were measured at concentrations ranging from 0.18 to 1.4 pg/L. Fish samples contained six congeners that included BDE-47, BDE-66, BDE-99, BDE-100, BDE-153, and BDE-154, with the highest values reported for lake trout collected in Lake Michigan, followed by Lakes Ontario, Superior, and Huron. Log BAFs for the PBDE congeners ranged from 6.7 to 7.5 for lake trout.

The levels and patterns of PBDE congeners were evaluated by Rice et al. (2002) in carp and largemouth bass collected from the Detroit River, MI, and in carp collected at two locations in the Des Plaines River, IL, in May and June 1999. Both are major rivers that receive industrial and municipal effluents. The average total concentrations of PBDEs in carp and bass from the Detroit River were 5.39 and 5.25 ng/g wwt, respectively. Higher total PBDE concentrations were reported for carp collected in the Des Plaines River, with an average total concentration of 12.48 ng/g wwt. The tetrabromo congener, BDE-47, represented 53 and 56% of the relative proportion of the seven detected congeners in carp and largemouth bass, respectively; these fish were collected from the Detroit River. In carp collected from the Des Plaines River, the most prominent congeners were BDE-183 and BDE-181 (20 and 22%, respectively, of the total PBDE residue), whereas the heptabromo congener BDE-190 was present in Des Plaines River samples and represented 12% of the total amount of PBDE. The authors also reported positive correlations between PBDE congener concentrations and increasing lipid levels, in both fish species. The authors suggested that the differences in dominant congener types between the two locations were probably related to differences in industrial and municipal discharges entering these systems.

Valters et al. (2005) assessed the concentrations of PBDEs and their metabolites, which included hydroxylated-PBDEs (OH-PBDEs) and methoxylated-PBDEs (MeO-PBDEs), in benthic and pelagic fish taken from the Detroit River collected in 2001 and 2002. Several other compounds were included in the study, including triclosan and its metabolites (methylated triclosan analogue). The dominant PBDE congeners were BDE-47, BDE-99, and BDE-100, and these accounted for 85% of the ΣPBDE concentration and individually ranged from 0.155 to 21.1 ng/g wwt. For comparison, triclosan concentrations ranged from 0.750 to >100 ng/g wwt. The dominant hydroxylated-PBDE was 6-OH-BDE-47 and had ΣOH-PBDE concentrations ranging from 0.003 to 0.198 ng/g wwt. MeO-PBDEs were below detection limits in all samples. The resulting ΣPBDE to ΣOH-PBDE ratio ranged from 0.0005 to 0.02. The authors suggested that OH-PBDEs are likely derived as metabolites from PBDE precursors and are retained in the blood.

The herring gull (*Larus argentatus*) is a good indicator organism for studying the bioaccumulation of organic compounds in ecosystems, because it fills the niche of facultative piscivore. Norstrom et al. (2002) studied the geographical distribution of PBDE flame retardants in herring gulls in 2002 by analyzing eggs collected at 15 locations in the Great Lakes. Temporal trends for the period from 1981 to 2000 were explored by analysis of archived egg homogenates from gull colonies in Lakes Michigan, Huron, and Ontario. A total of 25 di-BDE to hepta-BDE congeners were identified, with 7 congeners representing 97.5% of the ΣPBDEs. The dominant congener distributed throughout the Great Lakes was BDE-47, followed by BDE-99. Concentrations of ΣPBDE ranged from 192 to 1,400 ng/g wwt. The two highest concentrations were measured in gull colonies located in Lake Michigan (1,400 and 1,366 ng/g wwt), followed by Toronto Harbor in Lake Ontario (1,003 ng/g wwt). The lowest ΣPBDE concentrations were found in colonies in Lake Erie (192 ng/g wwt). An increase in ΣPBDE concentrations ranging from 20-fold to

75-fold occurred from 1981 to 2000 in herring gull eggs taken from colonies in Lake Michigan, Lake Huron, and Lake Ontario. The three most common congeners, BDE-47, BDE-99, and BDE-100, ranged from 5 to 12 ng/g wwt in 1981–1983, and then increased to 400–1,100 ng/g wwt from 1983 to 2000. The doubling times of these three congeners ($\Sigma PBDE_{47,99,100}$) were 2.8 years in Lake Ontario, 2.6 years in Lake Michigan, and 3.1 years in Lake Huron. The authors concluded that, if the present rate of change continued, the concentrations of ΣPBDE would equal or surpass those of ΣPCBs in Great Lakes herring gull eggs in 10–15 years.

Herring gull eggs were examined by Gauthier et al. (2007) for the presence of a variety of PBDE congeners (hepta-BDE to nona-BDE and BDE-209), including those that were not reported in spatial and temporal assessments of PBDEs from 1981 to 2000. The study included the analysis of other brominated and chlorinated flame retardants. BDE-209 and seven other heptabromo-BDE to nonabromo-BDE congeners were detected. The authors reported a trend of decreasing concentrations of $\Sigma_7$PBDEs when compared to data collected in 2000 (Norstrom et al. 2002). Of the other flame retardants that were included in the analysis, six were quantifiable in the egg samples pooled from all six colonies in 2004.

Gauthier et al. (2008) continued to utilize herring gull eggs to analyze the temporal trends of PBDEs, specifically BDE-209 in pooled samples from seven gull colonies in the Laurentian Great Lakes from 1982–2006. The authors reported that 39 congeners were quantifiable in the pooled egg samples, and the highest $\Sigma_{39}$PBDE concentration was detected in samples from Gull Island (1,191 ng/g wwt). In 2006, BDE-209 concentrations ranged from 4.5 to 20 ng/g wwt and constituted 0.6–4.5% of the $\Sigma_{39}$PBDE concentrations. The doubling time for BDE-209 between 1982 and 2006 was determined to be 2.1–3.0 years. Concentrations of $\Sigma_{octa}$BDEs and $\Sigma_{nona}$BDEs congeners detected in herring gull eggs may have resulted from the debromination of BDE-209 formed metabolically, either in the herring gulls or in the environment, and then accumulated through the diet and transferred to their eggs. The authors determined that congeners derived from the penta-PBDE and octa-PBDE mixtures (BDE-47, BDE-99, BDE-100) showed increases in concentrations up until 2000 with no increasing trend thereafter.

To assess sediment concentrations of various flame retardants, Qiu et al. (2007) reported the analysis of Dechlorane Plus (DP), PBDEs, and 1,2 bis(2,4, 6-tribromophenoxy)ethane (TBE) in sediment cores taken from the central basin of Lake Ontario. Of the PBDE concentrations detected, $\Sigma PBDE_{3-7}$ (sum of tri-BDE to hepta-BDE congeners) and BDE-209 were found at average concentrations of 2.8 and 14 ng/g dwt, respectively. BDE-209 was the dominant PBDE congener measured in sediment samples. The analysis of concentration profiles with depth suggested that BDE-209 accumulation in sediments increased beginning in the early 1980s.

Sediment core samples were taken from each of the Great Lakes in 2001–2002 and analyzed for the presence of PBDEs and PCBs. The results were published in a series of three papers by Song et al. (2004, 2005a, b). In the first paper, Song et al. (2004) sampled 6 locations in Lake Superior in 2001 and 2002 and reported the results for 10 PBDEs and 19 PCBs. $\Sigma PBDE_9$ (sum of nine congeners, excluding

BDE-209) concentrations ranged from 0.49 to 3.1 ng/g dwt in surficial sediments. ΣPCB (sum of all 19 PCB congeners) concentrations in surficial sediments were in the range of 2–27 ng/g dwt. BDE-209 was the predominant PBDE congener and represented 83–94% of the total concentration of PBDEs detected in the sediments. The authors estimated that the loading rate of all ten PBDE congeners to Lake Superior sediments was 80–160 kg/yr.

In subsequent work, sediment cores were collected at 6 sites in Lakes Michigan and Huron in 2002 and analyzed for the same 10 PBDE congeners, as well as 39 PCB congeners (Song et al. 2005b). The concentration of $\Sigma PBDE_9$ in surface sediments of Lake Michigan ranged from 1.7 to 4 ng/g dwt, whereas Lake Huron sediments contained $\Sigma PBDE_9$ congeners that ranged from 1.0 to 1.9 ng/g dwt. Sediment concentrations of PBDE showed a trend of increasing levels with decreasing depth. BDE-209 was the dominant congener representing 96 and 91% (by mass) of the PBDEs found in Lake Michigan (average of 63 ng/g dwt) and Lake Huron (average of 15.9 ng/g dwt), respectively. The total load of PBDEs in Lake Michigan, including BDE-209, was estimated to range from 29,000 to 50,000 kg with a total loading of 390–1,200 kg/yr. For Lake Huron, the total load was estimated to range from 6,000 to 15,000 kg (including BDE-209) and the loading rate from 400 to 840 kg/yr. Of the lower congeners, BDE-47 and BDE-99 were the most abundant. PCB concentrations were shown to be stable at most locations rather than continually increasing with decreasing sediment depth. Variations in the PCB data suggested that potential non-point source inputs of PCBs remain, although production and usage of PCBs were banned in the 1970s.

In 2002, sediment core samples were collected for evaluation of PBDE concentrations in Lakes Ontario and Erie (Song et al. 2005a). Sediment cores were obtained from 4 locations and analyzed for the same 10 PBDE congeners and 39 PCB congeners. $\Sigma PBDE_9$ concentrations (excluding BDE-209) ranged from 4.85 and 6.33 ng/g dwt, in Lake Ontario and from 1.83 and 1.95 ng/g dwt, in Lake Erie, respectively. BDE-209 concentrations were higher in Lake Ontario at 242 and 211 ng/g dwt, compared to 50 and 55 ng/g dwt, in Lake Erie. In Lake Ontario, PBDE concentrations followed a clear increasing trend from the bottom of sediment cores to the surface, where maximum concentrations were detected. In Lake Erie, the trends were not as clear, although the data were consistent with increasing concentrations of PBDE in surface sediments. The authors hypothesized that resuspension of sediments occurs in the shallow waters of Lake Erie from the action of wind and waves. The observation that PBDE concentration increases with decreasing sediment depth supports the view that there has been a continuous input of PBDEs since the 1970s. Although average PBDE concentrations were reported to be higher for Lake Ontario than for Lake Erie, organic carbon normalized data were similar between the lakes. The current load of PBDEs, including BDE-209, was 22 and 18 t for Lake Ontario and Lake Erie, respectively, in 2002. Loading rates were determined to be 3,400 kg/yr in Lake Ontario for total PBDEs (all ten congeners) and 32 kg/yr and 1,311 kg/yr for $\Sigma PBDE_9$ and BDE-209, respectively, in Lake Erie. BDE-209 was the dominant PBDE congener measured in sediments in both lakes, and respectively, represented 91 and 96% of the measured total PBDE concentrations in Lakes Erie and Ontario. Of the remaining nine PBDE congeners, BDE-47 and BDE-99 were the most common. PCB concentrations peaked in the 1960s and 1970s and showed a trend of decreasing levels in surface sediments. In 2002, sediment cores from Lake Ontario exhibited surface PCB concentrations ranging from 58.3 to 63.6 ng/g dwt, whereas Lake Erie had lower concentrations of 23–28.3 ng/g dwt.

A database of 2,197 data points was created from 19 published studies in which the concentrations of PBDEs in the various environmental media were evaluated. Although concentrations have been reported for as many as 28 different congeners, groups of congeners, or metabolites, to facilitate the interpretation, the analytes were first grouped into categories on the basis of data availability. For the present analysis, metabolite data were not included. When possible, individual BDE congeners were grouped to derive concentrations representative of the commercial products (penta-PBDE, octa-PBDE, or deca-PBDE). For the herring gull egg data, individual congeners were grouped according to the degree of bromination (i.e., total $Br_3$, total $Br_4$, total $Br_5$, total $Br_6$, etc.). As presented in Table 13a and b, concentrations in biological tissues ($n = 673$) represented the majority of the data; fewer data were available for sediment ($n = 45$).

Table 13a presents the analysis of concentrations of PBDEs in fish. Note that, for this dataset, concentrations of individual congeners were reported, and these were grouped into categories representative of the commercial products. Penta-PBDE concentrations in fish ranged from 0.005 to 354.7 ng/g wwt; mean and 95th percentile values were 35.1 and 122.7 ng/g wwt, respectively. Concentrations of octa-PBDE detected in fish were lower, ranging from 0.77 to 1.92 ng/g wwt. Mean and 95th percentile octa-PBDE concentrations were 1.45 and 1.81 ng/g wwt, respectively.

The most abundant single congener detected in fish was BDE-47, followed by BDE-99 and BDE-100 (Fig. 9a and b). The predominance of congener data available for commercial penta-PBDE is evident in Table 13a, with 235 values as compared to 22 for the octa-PBDE product. The higher concentrations of these lower-brominated congeners may reflect the increased production and use of the penta-PBDE products or differences in bioavailability.

In Table 13a, we also summarize the analysis of concentrations of various congener groups in herring gull eggs. Of the various congeners, total $Br_4$ BDE and total $Br_5$ BDE were predominant. Concentrations for these groups ranged from 0.06 to 708 ng/g wwt; mean and 95th percentile ranged from 106 to 437 ng/g wwt. The prevalence of $Br_4$ and $Br_5$ congeners in the gull eggs is consistent with the distribution of PBDE congeners in fish. Although fewer data were available, concentrations of the higher brominated congeners were lower in the gull eggs, reflecting either differences in production and use or decreased bioavailability of the higher congeners.

As discussed above, numerous researchers have documented trends of increasing PBDE concentrations in biota dating back to 1979 and continuing through 2000. The data suggested that PBDEs initially began to accumulate in biota around 1980, which corresponds to increasing production and use of these flame retardants.

**Table 13** Summary Statistics for Polybrominated diphenyl ethers (PBDE)

| | *n* | Freq det (%) | Mean (ng/g wwt) | SD (ng/g wwt) | Median (ng/g wwt) | 95th percentile (ng/g wwt) | Min (ng/g wwt) | Max (ng/g wwt) |
|---|---|---|---|---|---|---|---|---|
| *A. Biota* | | | | | | | | |
| Fish | | | | | | | | |
| Penta[1] | 235 | NA | 35.1 | 54.1 | 10.4 | 122.7 | 0.005 | 354.7 |
| Octa | 22 | NA | 1.45 | 0.299 | 1.53 | 1.81 | 0.77 | 1.92 |
| Gull eggs | *n* | Freq det (%) | Mean (ng/g wwt) | SD (ng/g wwt) | Median (ng/g wwt) | 95th percentile (ng/g wwt) | Min (ng/g wwt) | Max (ng/g wwt) |
| Total $Br_3$ BDE | 55 | NA | 3.03 | 2.34 | 2.7 | 7.03 | 0.1 | 8.2 |
| Total $Br_4$ BDE | 55 | NA | 171 | 144 | 128 | 437 | 2.8 | 602 |
| Total $Br_5$ BDE | 103 | NA | 106 | 128 | 63.4 | 364 | 0.6 | 708 |
| Total $Br_6$ BDE | 103 | NA | 44.9 | 45.2 | 31.8 | 142 | 1.4 | 235 |
| Total $Br_7$ BDE | 61 | NA | 11.9 | 18.8 | 7.1 | 43.5 | 0.2 | 125 |
| Total $Br_8$ BDE | 13 | NA | 6.85 | 4.67 | 5.4 | 15.4 | 2.6 | 19 |
| Total $Br_9$ BDE | 13 | NA | 2.65 | 2.48 | 2.3 | 6.78 | 0.1 | 8.5 |
| BDE-209 | 13 | NA | 6.35 | 7.31 | 4.5 | 18.8 | 0.05 | 20 |
| | *n* | Freq det (%) | Mean (ng/g dwt) | SD (ng/g dwt) | Median (ng/g wwt) | 95th percentile (ng/g dwt) | Min (ng/g dwt) | Max (ng/g dwt) |
| *B. Sediment* | | | | | | | | |
| Penta PBDE[2] | 22 | 100 | 2.23 | 1.42 | 1.83 | 4.81 | 0.49 | 6.33 |
| Deca PBDE | 23 | 100 | 71.3 | 88.5 | 36 | 239 | 4.30 | 315 |

[1]Penta-PBDE and octa-PBDE concentrations were calculated by summing values for brominated diphenyl ether (BDE) isomer BDE-47 through BDE-154 and BDE-153 through BDE-190, respectively

[2]Penta-PBDE concentrations were calculated by summing values for BDE-28 through BDE-183

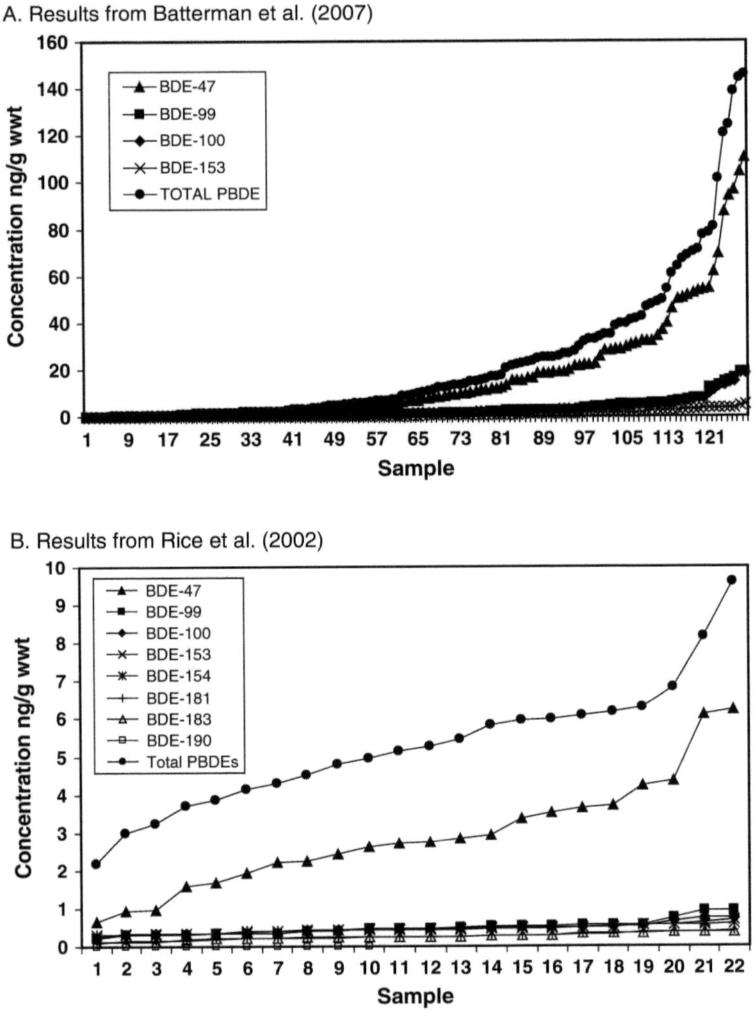


**Fig. 9** Distribution of BDE congeners in fish compared with total PBDE concentrations. **a** Results from Batterman et al. (2007) and **b** results from Rice et al. (2002)

However, these data are biased, because few if any pre-1980 archived biota samples are available for analyses. Nonetheless, the trend of increasing PBDE concentrations through the 1980s into the 1990s is consistent with the findings of Batterman et al. (2007), Chernyak et al. (2005), and Zhu and Hites (2004). More recent data (post-2000) suggest that concentrations in biota have leveled-off or are declining, particularly for penta-PBDE and octa-PBDE, which likely reflects changes in use of these products as a consequence of regulatory bans or, for some products, voluntary decisions to stop manufacture.

Deca-PBDE (BDE-209) was not reported in any fish samples from the Great Lakes. Possible explanations include the exclusion of this specific congener from analyses, the limited manufacture and release of BDE-209 in the early to mid-1980s, or the reduced bioavailability and bioaccumulation potential of BDE-209 in fish. However, several researchers have noted that a potential source of hepta-BDE congeners (BDE-181, BDE-183, and BDE-190) may be from the environmental transformation of BDE-209. Some researchers have suggested that the debromination of deca-PBDE may result in the formation of hepta-BDE and octa-BDE congeners, which become more bioavailable to accumulate in fish and other organisms.

Norstrom et al. (2002) performed analysis of herring gull eggs collected from 1981 to 2000 and reported that nona-BDEs and BDE-209 were not detected in eggs collected from 15 locations around the Great Lakes. However, low concentrations of BDE-209 were detected at quantifiable levels in some samples that were collected at the same locations in 2004 (Gauthier et al. 2007). From the statistical analysis, mean concentrations of BDE-209 in gull eggs were 6.35 $\pm$ 7.31 ng/g wwt ($n = 13$), with minimum and maximum concentrations of 0.05–20 ng/g wwt, respectively. These values were similar to the 5th and 95th percentile values (0.05 and 18.8 ng/g wwt).

Although the data are more limited, penta-PBDE and deca-PBDE have been detected in sediment cores, and concentrations in such cores appear to be increasing over time. As summarized in Table 13b, penta-PBDE concentrations range from 0.049 to 6.33 ng/g dwt; mean and 95th percentile concentrations were 2.23 and 4.81 ng/g dwt, respectively. In contrast, BDE-209 was the dominant congener detected, representing over 85% of the PBDEs analyzed in sediment samples. The BDE-209 concentrations ranged from 4.3 to 315 ng/g dwt (mean and 95th percentile of 71.3 and 239 ng/g dwt, respectively). Increasing trends in sediment deca-PBDE concentrations may be due to the increased production of deca-PBDE in the late 1990s combined with the propensity to partition to sediment particles because of having a high $K_{OW}$.

Risk assessments for commercial PBDEs have been conducted by authorities in various geographies (EC, 2006b; EUC 2001, 2003b). Estimated no effect values and PNECs have been developed for use in the calculation of risk quotients (Table 2). Comparison of the available data for sediment concentrations (Table 13b) versus the various regulatory criteria suggests that all exposures are below effect concentrations developed in Canada and Europe. Maximum penta-PBDE and deca-PBDE concentrations in sediment are 0.006 and 0.315 mg/kg dwt, respectively, which are below the most stringent criteria. Maximum and 95th percentile values for penta-PBDE in fish are 0.35 and 0.12 mg/kg wwt, respectively (Table 13a). These concentrations exceed the Canadian ENEV for secondary poisoning via the food chain, although they are below the PNEC value derived in Europe (Table 2). In contrast, all octa-PBDE concentrations in fish (maximum = 0.002 mg/kg food) are below the regulatory criteria.

The lower brominated congeners were more prevalent in gull egg samples, with the tetra-BDE and penta-BDE congeners being present at the highest concentrations. Mean values were 0.17 and 0.11 mg/kg wwt, and the corresponding maximum concentrations were 0.6 and 0.71 mg/kg wwt, respectively. These concentrations exceed the Canadian ENEV criteria value for secondary consumers (0.0084 mg/kg food). The remaining PBDE congeners reported for gull eggs were below the regulatory criteria for both Canada and Europe, including BDE-209 concentrations, which had mean and maximum values of 0.006 and 0.02 mg/kg wwt, respectively.

## *3.8 Other Flame Retardants*

The use of flame retardants in commercial and consumer goods has grown over the last 40 years as a result of increasingly stringent fire safety regulations. As discussed in the previous section, PBDE flame retardants were widely used for many years in a variety of applications. However, because of increased environmental and human health concerns, several of the PBDEs (e.g., penta-BDE and octa-BDE) have been recently banned in Europe and in other countries. The major manufacturer in the United States stopped production of these compounds in 2004. As flame retardants are taken off the market, they are replaced by others to enable manufacturers of commercial and consumer goods to continue to meet government safety standards.

The substances discussed in this category include a variety of other flame retardants, many of which have been produced for more than 40 years. As presented below, environmental concentrations have been reported for a variety of these substances, particularly for Dechlorane Plus (DP) and hexabromocyclododecane (HBCD) in sediments and biota. Limited information is available for various other substances and for other environmental media.

In the present analysis we focused on those studies that reported concentrations in surface waters, sediment, and biota (sampling locations not illustrated). However, a number of studies that were identified in the literature included analytical results for flame retardants in air (e.g., Hoh and Hites 2005; Hoh et al. 2005, 2006; Venier and Hites 2008).

Hoh et al. (2005) analyzed for TBE and pentabromoethylbenzene (PBEB) in a single sediment core taken from northern Lake Michigan in April 2004. TBE was detected in the sediment core from Lake Michigan at about 9 ng/g dwt, whereas PBEB was not detected. Based on the analysis of sediment samples taken at various depths, TBE first appeared in the core at a depth corresponding to the year 1973, consistent with the beginning of production. Levels in the core increased toward the shallower depths with a doubling time of approximately 2 years until 1985, when concentrations stabilized. TBE was not detected in the surficial (top) layer of the core which represented 1993–2004. According to US EPA Inventory Update Rule, TBE production increased from 1986 to 1994 and then decreased substantially after 1998.

Hoh et al. (2006) summarized the analysis of DP in sediment and fish collected throughout the Great Lakes Basin. DP is a chlorinated flame retardant introduced in the mid-1960s as a substitute for Dechlorane (or Mirex) and is incorporated into polymers used for coating electrical wires and cables, connectors used in computers, and plastic roofing material. Two sediment cores were taken from eastern Lake Erie in August 2003, and one each was taken from northern and southern Lake Michigan in April 2004. Archived fish samples, taken from Lake Erie throughout 1980–2000 (seven sampling periods), were also analyzed. DP was detected in sediment cores taken from Lakes Michigan and Erie. Concentrations in the cores appeared to peak during the period between 1976 and 1981, and since then have declined by about 50% in surface layers. The highest historical DP concentration in Lake Erie (40 ng/g dwt) was about ten times higher than observed in the sediment core profiles for Lake Michigan. Concentrations in the surface layers of Lakes Michigan and Erie sediments were in the range of 2–5 ng/g dwt.

Because of the higher concentrations of DP in air and sediments from Lake Erie, Qiu et al. (2007) speculated that Lake Ontario may receive atmospheric contributions of DP from a manufacturing facility in Niagara Falls, NY. To address this, a sediment core was obtained in 2004 from the central basin of Lake Ontario and analyzed for DP and other brominated flame retardants. Results showed that DP concentrations increased rapidly in the mid-1970s and reached a peak concentration (310 ng/g dwt) in the mid-1990s. DP concentrations near the surface were lower (150 ng/g dwt). Measurements of TBE in the sediment core showed increased levels starting after the early 1980s, with a maximum concentration of 6.7 ng/g dwt in the surficial sediment.

Sverko et al. (2008) reported the results of a survey of DP concentrations in sediments collected during the period from 1997 to 1998 from 40 locations in Lakes Erie and Ontario. Archived samples of Niagara River suspended sediments were also analyzed. Total DP concentration in surface sediments from Lakes Erie and Ontario ranged from 0.061 to 8.62 ng/g and from 2.23 to 586 ng/g dwt, respectively. Lake-wide average DP concentrations in Lake Ontario were about 50 times greater than observed in Lake Erie. Analysis of archived sediments collected from the Niagara River, during the period from 1980 to 2002, showed a declining total DP concentration from 89 to 7 ng/g dwt, suggesting a possible decrease in production or the reduction of DP released to the environment from the manufacturing facility. The spatial distribution of DP in Lake Ontario was generally related to the bathymetry, with the highest concentrations associated with fine-grained sediments in the three major deep-water depositional basins. The non-depositional sill zones, where the sediments were characterized by coarse sands, exhibited the lowest levels. Although the Niagara River has been suggested as a primary source of DP, the spatial distribution of DP was noted as being similar to that of other contaminants, including PCBs. The highest concentration measured in Lake Ontario was near the city of Toronto, which is unusual if one considers the river as the primary source and the counterclockwise flow pattern of the lake. The authors concluded that in addition to the Niagara River, other sources may also be important contributors to DP concentrations in Lake Ontario, and that greater spatial resolution is required.

The distribution of HBCD isomers in Detroit River sediments was reported by Marvin et al. (2006). HBCD is the principal flame retardant in extruded and expanded polystyrene foams and is used as insulation in the building industry. The substance is also used in upholstery textiles, including residential and commercial furniture, draperies, and wall coverings. Suspended sediment samples ($n = 63$) were collected monthly (from May to November, 2001) from the Detroit River from single-point sediment-trap moorings sited at nine stations, located from the headwaters in southern Lake St. Clair to the mouth of the outflow to western Lake Erie. Suspended sediments were sampled instead of bottom sediments because of the non-depositional nature of the upper and middle reaches of the river. Individual HBCD isomers were found at relatively low concentrations ranging from < 0.025 to 1.9 ng/g dwt for the alpha isomer, to < 0.025 to 2.3 ng/g dwt for the gamma isomer; concentrations of the beta isomer were lower (< 0.025–0.28 ng/g dwt). Approximately 66% of the suspended sediment samples were dominated by the gamma isomer in which the isomeric profiles were similar to the commercial technical mixture. The isomer profiles in the remaining samples were dominated by the alpha isomer. The highest monthly concentrations of total HBCD were detected in samples from the upper reaches of the river near Belle Isle (2.6 ng/g dwt, in October), near the mouth of the Rouge River (3.65 ng/g dwt, in August), and near the head of the Trenton Channel (2.6 ng/g dwt, in August). The authors noted that the distribution and occurrence of HBCD in suspended sediments of the Detroit river were commensurate with land use patterns (i.e., general urbanization and industrialization) and do not provide evidence of the presence of significant point sources. The widespread occurrence of relatively low concentrations of HBCD in the sediments suggests that large urban areas can act as diffuse sources.

Tomy et al. (2004) analyzed the bioaccumulation and trophic magnification of HBCD isomers in a Lake Ontario pelagic food web. Lake trout ($n = 5$), alewife (composites of 5 fish, $n = 3$), rainbow smelt (composites of 5 or 20 fish, $n = 3$), and slimy sculpin (composites of 10 or 15 fish, $n = 3$), were collected between June and September of 2002 at offshore stations in Lake Ontario. Invertebrate species, including samples of *Mysis* (composites of >100, $n = 2$) and amphipods (*Diporeia*, composites of >100, $n = 2$), as well as plankton grab samples ($n = 2$), were collected at the same locations. Concentrations of the alpha isomer were consistently higher than that of the gamma isomer, while levels of the beta isomer were always below the detection limit. Measured concentrations of alpha (0.4–3.8 ng/g wwt) and gamma HBCD (0.1–0.8 ng/g wwt) were highest in the lake trout samples. For the prey fish species, trends in alpha and gamma levels were on the order of slimy sculpin > smelt > alewife. Because of their benthic association and higher lipid content, it is reasonable that levels of HBCD were higher in sculpin compared to the other two forage fish species. Mean concentrations in *Mysis* and *Diporeia* were similar (0.14–0.16 ng/g) and about twofold higher than concentrations measured in plankton (0.06 ng/g wwt). Biomagnification factors were variable between the trophic levels and feeding relationships and ranged from 0.2 to 10.8. The trophic magnification factor (TMF) of 6.3, derived for HBCD, is comparable to those reported for DDE (6.1) and the PCBs (5.7).

In a related study, Tomy et al. (2007) examined the bioaccumulation and trophic magnification of DP in a Lake Ontario pelagic food web. Biological specimens consisting of plankton, invertebrates, forage fish, and lake trout were collected between June and September of 2002, using the sampling protocols discussed above. Concentrations of the *anti*-isomer were generally higher than that of the *syn*-isomer.

The extent of bioaccumulation of the isomers was greater in the lower trophic levels in Lake Ontario. For example, measured concentrations of *anti*-DP and *syn*-DP were the highest in *Diporeia* (3.1 ± 0.9 ng/g lipid wt and 1.3 ± 0.6 ng/g lipid wt, respectively) and were approximately 10–30-fold lower in lake trout. For the prey fish species, *anti*-isomer and *syn*-isomer concentrations ranged from 7 to 901 ng/g lipid wt, and the trends in total DP concentrations were on the order of slimy sculpin > alewife > smelt. Biomagnification factors were variable between the trophic levels and feeding relationships and ranged from 0.1 to 12. However, no statistically significant trophic magnification factor for either isomer was determined for the Lake Ontario food web. In contrast, related work conducted in Lake Winnipeg resulted in trophic magnification factors of 2.5 and <1 for the *anti*-isomer and *syn*-isomer, respectively. Further analysis of the results suggested that interspecies differences in bioaccumulation and biotransformation are probable key factors in determining the distribution of DP isomers in the environment.

Gauthier et al. (2007) analyzed for the presence of a variety of flame retardants in gull eggs that were collected throughout the Laurentian Great Lakes. Eggs ($n = 10$–13) were collected from six colonies in April–May of 2004, and pooled homogenates were prepared and analyzed for each colony. As illustrated in Fig. 10, concentrations of the flame retardants, including PBEB, DP, and HBCD, ranged among the colonies from 0.03 to 1.4 ng/g wwt, from 1.5 to 4.5 ng/g wwt, and from <0.01 to 20.7 ng/g wwt, respectively. Concentrations of other flame retardants, including hexabromobenzene (HBB), pentabromotoluene (PBT), and 1,2-bis(2,4, 6-tribromophenoxy)ethane (BTBPE or TBE) were all less than 1 ng/g wwt. Whether the latter substances are still used commercially was unclear. In contrast, total PBDE

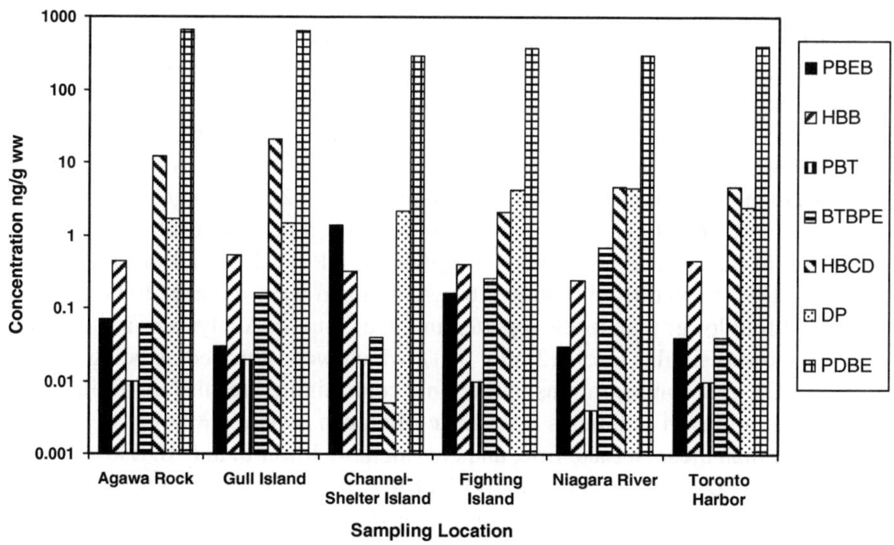


**Fig. 10** Distribution of flame retardants in gull eggs from the Laurentian Great Lakes

concentrations in the eggs from all colonies were substantially higher, ranging from 293 to 663 ng/g wwt. The most abundant PBDE congeners detected were those associated with penta-PBDE mixtures. Concentrations of total hepta-BDE, octa-BDE, and nona-BDE were generally less than 10 ng/g wwt, whereas deca-BDE was typically below the detection limit. When PBDE concentrations measured in the current study were compared with those determined in 2000 for eggs collected from the same colonies, the trend appears to suggest that concentrations have decreased considerably. The limited data precludes an analysis of trends in concentrations for the remaining compounds. The authors concluded that mother herring gulls are exposed to several current-use flame retardants via their diet, and that in ovo transfer occurs to their eggs.

Only limited information was available for the presence of several chlorinated phosphate flame retardants in surface waters of the Great Lakes. Kolpin et al. (2002) reported concentrations in water samples ($n = 7$) from the Lake Michigan watershed that ranged from below the detection limit to 0.27 μg/L for tri-(2-chloroethyl)phosphate and tri-(dichloroisopropyl)phosphate. Andresen et al. (2007) reported concentrations for these two as well; tri-(dichloropropyl) phosphate ranged from 0.002 to 0.08 μg/L in samples ($n = 6$) from Hamilton Harbor, ON. According to the authors, these compounds were previously used as flame retardants in polyurethane foam and have been phased out because of toxicity concerns.

The statistical analysis of the flame retardant data in this category was performed on a total of 660 data points. As summarized in Table 14a–c, the concentrations reported in biological tissues ($n = 509$) represented the majority of the data. Fewer data were available for sediments ($n = 119$) and only for a few of the compounds. Limited data were available for surface waters.

Concentrations of DP have been reported for sediment and biological samples collected throughout the Great Lakes Basin. Several of the studies focused on the potential contributions of a manufacturing facility located in the Niagara River region. Information on the concentrations and spatial distribution of DP in sediments was available from three studies, although the majority of samples were obtained from Lakes Erie and Ontario. A survey conducted in 1997 and 1998 showed that DP levels were the highest in Lake Ontario, ranging from 2.2 to 586 ng/g dwt. Analysis of a single core from Lake Ontario in 2003 showed comparable results; the DP concentration in the surface layer was 150 ng/g dwt. DP levels found in Lake Erie in 1997 and 1998 ranged from 0.06 to 8.6 ng/g dwt and were reported to be approximately 50-fold lower than levels reported in Lake Ontario. Analysis of a single core in 2003 showed similar levels (3 ng/g dwt), which were reported as being comparable to BDE-209 concentrations in the same core. Limited results are available for DP concentrations in sediments from Lake Michigan, where the levels were found to be lower than those in Lake Erie, and considerably lower than PBDE levels in the same samples. DP was detected in 100% of the aquatic biota collected from Lake Ontario (Table 14c). The concentrations ranged from 0.02 to 4.4 ng/g wwt (mean $= 1.27 \pm 1.59$ ng/g wwt) and were highest in *Diporeia*. Lower levels in lake trout (approximately 0.02 ng/g wwt) were consistent with decreased bioaccumulation of

**Table 14** Summary statistics for other flame retardants

| | *n* | Freq det (%) | Mean (μg/L) | SD (μg/L) | Median (μg/L) | 95th percentile (μg/L) | Min (μg/L) | Max (μg/L) |
|---|---|---|---|---|---|---|---|---|
| *A. Surface water* | | | | | | | | |
| Tri (2-chloro-ethyl) phosphate | 13 | 85 | 0.086 | 0.095 | 0.035 | 0.240 | 0.004 | 0.270 |
| Tri (di-chloro-isopropyl) phosphate | 13 | 62 | 0.052 | 0.033 | 0.050 | 0.104 | 0.003 | 0.110 |
| Tri(dichloro-propyl phosphate) | 6 | 100 | 0.015 | 0.014 | 0.011 | 0.033 | 0.002 | 0.035 |
| | *n* | Freq det (%) | Mean (ng/g dwt) | SD (ng/g dwt) | Median (ng/g dwt) | 95th percentile (ng/g dwt) | Min (ng/g dwt) | Max (ng/g dwt) |
| *B. Sediment* | | | | | | | | |
| Total Dechlorane Plus (DP) | 48 | 100 | 107.8323 | 217.7972 | 5.0000 | 455.2000 | 0.2060 | 586 |
| Total HBCD | 66 | 64 | 0.4702 | 0.7273 | 0.1825 | 2.2275 | 0.0008 | 3.65 |
| Pentabromoethyl benzene (PBEB) | 1 | 0 | 0.01 | 0.00 | 0.01 | 0.01 | 0.01 | 0.01 |
| Tribromophenoxy ethane (TBE) | 4 | 75 | 5.27 | 4.62 | 6.70 | 8.77 | 0.10 | 9.00 |

**Table 14** (continued)

| | *n* | Freq det (%) | Mean (ng/g wwt) | SD (ng/g wwt) | Median (ng/g wwt) | 95th percentile (ng/g wwt) | Min (ng/g wwt) | Max (ng/g wwt) |
|---|---|---|---|---|---|---|---|---|
| *C. Biota*[a] | | | | | | | | |
| Aquatic | | | | | | | | |
| Total DP | 21 | 100 | 1.27 | 1.59 | 0.50 | 3.71 | 0.02 | 4.42 |
| Total HBCD | 20 | 100 | 0.56 | 1.01 | 0.20 | 1.57 | 0.02 | 4.51 |
| Gull Eggs | *n* | Freq det (%) | Mean (ng/g wwt) | SD (ng/g wwt) | Median (ng/g wwt) | 95th percentile (ng/g wwt) | Min (ng/g wwt) | Max (ng/g wwt) |
| Bis-tribromophenoxy ethane (BTBPE) | 78 | 100 | 0.210 | 0.255 | 0.110 | 0.590 | 0.040 | 0.700 |
| Total DP | 78 | 100 | 2.77 | 1.28 | 2.35 | 4.43 | 1.50 | 4.50 |
| Hexabromobiphenyl (HBB) | 78 | 100 | 0.400 | 0.104 | 0.425 | 0.513 | 0.240 | 0.530 |
| Total HBCD | 78 | 83 | 7.39 | 7.71 | 4.66 | 18.56 | 0.01 | 20.67 |
| PBEB | 78 | 100 | 0.288 | 0.547 | 0.055 | 1.090 | 0.030 | 1.400 |
| Pentabromotoluene (PBT) | 78 | 100 | 0.012 | 0.006 | 0.010 | 0.020 | 0.004 | 0.020 |

[a] The results represent concentrations reported for multiple trophic levels (invertebrates, small fish, large fish)

DP at higher trophic levels that were reported for this lake. DP was also detected in 100% of the gull eggs collected throughout the basin. Concentrations ranged from 1.5 to 4.5 ng/g wwt (mean = 2.8 ± 1.3 ng/g wwt). The highest concentration in gull eggs was measured in samples collected from sites near the Niagara River.

Fewer data were available for the concentrations of HBCD in the Great Lakes Basin. In one study, sediment concentrations obtained from various locations along the Detroit River were analyzed. As shown in Table 14b, HBCD (total) was detected in 64% of these sediment samples at concentrations ranging from below the detection limit to 3.65 ng/g dwt (0.47 ± 0.73 ng/g dwt). The highest concentrations were present at the confluence of the Rouge and Detroit Rivers. The authors noted that HBCD concentrations in the Detroit River represent the contributions from diffuse sources and were three orders of magnitude lower than were PCB concentrations in this region. HBCD was detected in 100% of the aquatic biota collected from Lake Ontario. HBCD concentrations ranged from 0.02 to 4.5 ng/g wwt (mean 0.56 + 1.0 ng/g wwt) and increased in samples from the various trophic levels (invertebrates, prey fish, and large fish). The highest concentrations were reported in lake trout, consistent with the calculated trophic magnification factor of 6.3. HBCD was detected in 83% of the gull eggs collected from locations throughout the basin. Concentrations ranged from non-detectable to 20.7 ng/g wwt. The highest concentrations were measured from eggs collected from northern Lake Michigan (20.7 ng/g wwt) and Agawa Rock (12.2 ng/g wwt) in Lake Superior (Fig. 10). Elsewhere, HBCD concentrations ranged from 2.1 to 4.7 ng/g wwt and were not detectable in eggs collected from Channel Island in Saginaw Bay.

Only limited data were available for the remaining flame retardants examined in this category. Low concentrations of TBE (also referred to as BTBPE) were measured in sediments, ranging from 0.1 to 9 ng/g dwt. BTBPE was detected in 100% of the gull eggs at concentrations ranging from 0.04 to 0.7 ng/g wwt. PBEB was detected in all of the gull eggs sampled at concentrations ranging from 0.03 to 1.4 ng/g wwt and was not detected in the single sediment sample which was analyzed.

In contrast to the PBDEs, few risk assessments have been performed for the flame retardants discussed in this section. The European Chemicals Bureau of the EUC (EUC 2008 g) has recently completed a comprehensive risk assessment for HBCD. Although no risks were indicated for certain uses, risk quotients greater than one were concluded for others, including general uses in textile back coating and certain uses of expanded polystyrene. Further, although the data are equivocal and under debate, HBCD was considered to meet criteria for persistence, bioaccumulation, and toxicity (PBT), and an overall conclusion of risk was indicated for secondary poisoning. As a result, HBCD has recently been nominated for evaluation under the United Nations Economic Commission for Europe Long Range Transboundary Air Pollution Convention (UNECE LRTAP) and under the United Nations Environment Program (UNEP) Stockholm Convention. In the EU risk assessment, PNECs for HBCD are reported for surface waters, sediments, and for secondary poisoning by ingestion of food. These PNEC values are 0.31 μg/L, 0.86 mg/kg dwt, and 15.3 mg/kg food, respectively (Table 2). When compared to the limited data for the Great Lakes Basin, maximum reported concentrations of HBCD in sediments (0.0037 mg/kg dwt) and biota (0.02 mg/kg wwt) are well below these PNEC values.

### *3.9 Chlorinated Paraffins*

Chlorinated paraffins are complex mixtures of chlorinated alkanes produced since 1930 by direct chlorination of normal paraffins fractions. Commercial products are usually differentiated into three categories depending on chain length; short chain ($C_{10-13}$), medium chain ($C_{14-17}$) and long chain ($C_{18-30}$). Short-chain chlorinated paraffins (SCCP) are primarily used as extreme temperature additives in metal-working fluids, although they are also used in paints and sealants, and in the leather-working industry. Medium- (MCCP) and long-chain (LCCP) products are used as flame retardant plasticizers and as additives to improve the water resistance and flame retardancy of adhesives, paints, and sealants. SCCPs have attracted the greatest attention owing to their toxicity, persistence, and bioaccumulation potential. Environmental releases of these products may occur during production, storage, transportation, and use of chlorinated paraffin-based products, or from emissions from wastewater treatment plants, landfills, and other waste disposal sites.

Compared to monitoring data for other halogenated organics in the Great Lakes, the number of studies and the geographical distribution of data for chlorinated paraffins are rather limited. Factors contributing to the lack of measurements have been the availability of standards and the development of analytical techniques for these complex mixtures. The majority of the available data for water and sediment are for samples from Lake Ontario, although some surface water data are also available for the St. Lawrence River (sampling locations not illustrated). Limited data are available for the presence of SCCPs in biological tissues from Lakes Michigan and Ontario.

Muir et al. (2001) summarized the analytical results of surface waters, sediments, and biota collected from Lake Ontario. Surface sediments were collected from three harbor areas (Toronto, Port Credit, and Hamilton, ON), in 1996. Carp were also collected in 1996 from Hamilton Harbor, and lake trout were obtained from two locations in western Lake Ontario (Port Credit and Niagara on the Lake). Large volume samples of lake water were obtained in July 1999 from the western basin of Lake Ontario. SCCPs were present in the surface water samples; sum SCCP concentrations were 1.75 ng/L and were dominated by $C_{12}$ congeners. Analysis of atmospheric data and fugacity ratios suggested that $C_{10-12}$ homologs were volatilizing from water to air. SCCPs were measured in all surface-sediment samples from the three harbors. Sum SCCP concentrations in sediment ranged from 7.3 to 290 ng/g dwt. The highest concentrations were found at an industrialized location (Windermere basin in Hamilton Harbor), which has documented heavy metal, PAH, and PCB concentrations. SCCPs were detectable in all samples of carp and lake trout. Average concentrations in carp (sum SCCP $= 2{,}630 \pm 2{,}560$ ng/g wwt) were higher than those detected in lake trout ($59 \pm 51$–$73 \pm 47$ ng/g wwt); these values were consistent with the observation that SCCP congeners are metabolized and have short half-lives and low biomagnification factors in juvenile rainbow trout. For comparison, the mean sum PCB concentrations in the same trout ranged from 2,180 to 5,150 ng/g wwt.

Moore et al. (2004) described the development of a sophisticated GC/MS method for quantifying SCCP concentrations in environmental samples. The method was applied to the analysis of high volume water samples from the St. Lawrence River. Samples were collected in April, September, and October in 1999 from the drinking water intake of the municipality of Levis, QC. Concentrations of sum SCCP ranged from 3.48 to 38.44 ng/L. No apparent seasonal differences in concentrations were noted.

Marvin et al. (2003) reported the results of a survey of SCCP concentrations in sediments collected in 1998 from 27 stations in Lake Ontario. Selected stations in each of the depositional basins exhibited the highest concentrations, ranging from 147 to 410 ng/g dwt. The average sum SCCP concentration lakewide was 49 ng/g dwt. Because of the limited sampling frequency, the distribution of SCCPs did not exhibit a definitive spatial trend. Assessment of core profiles and estimates of SCCP fluxes indicated that an area in the western basin of the lake was heavily impacted and potentially influenced by local industrial sources. Maximum accumulation of SCCPs in the western basin occurred in the mid-1970s. In contrast, SCCP concentrations in cores from the central basin of the lake exhibited characteristics that were more similar to remote locations that were primarily impacted by atmospheric sources.

Houde et al. (2008) examined the distribution, bioaccumulation, and trophic biomagnification of SCCP and MCCP in Lake Ontario and northern Lake Michigan. Water samples ($n = 10$) were collected from three sites in Lake Ontario (western, central, and eastern basins) during the period from October 2000 to July 2004. Net plankton, forage fish, and lake trout ($n$ varies from 2 to 7) were collected in northern Lake Michigan (near Charlevoix, MI) and from locations in western Lake Ontario in July 2001. Total sum SCCP and sum MCCP concentrations in lake water from Lake Ontario were 1,190 and 0.9 pg/L, respectively. Despite different sampling techniques, different years, and different analytical methods, the total SCCP concentrations in the water samples from the two lakes were similar. Chlorinated paraffins were also detected in invertebrates and fish from both lakes. SCCP predominated in organisms from Lake Michigan, with the highest mean concentrations detected in lake trout ($123 \pm 35$ ng/g wwt). In contrast, MCCPs dominated most species in Lake Ontario, with the highest levels detected in slimy sculpin (108 ng/g wwt). Log bioaccumulation factors for lake trout ranged from 4.1 to 7.0 for SCCPs and from 6.3 to 6.8 for MCCPs. SCCP and MCCP were found to biomagnify between prey and predators from both lakes. Trophic magnification factors for invertebrates–forage fish–lake trout food webs ranged from 0.41 to 2.4 for SCCPs and from 0.06 to 0.36 for MCCPs.

Preliminary results for SCCP and MCCP concentrations in Lake Ontario biota have been recently reported by Tomy et al. (2007). The study focused primarily on the bioaccumulation of Dechlorane Plus in Lake Ontario food webs, but included the analysis of SCCPs and MCCPs in extracts from archived samples of invertebrates, forage fish, and lake trout collected between 2000 and 2003. Limited details were provided in the paper, and future publication of the data was indicated. Sum SCCP concentrations (normalized-to-lipid content) ranged from 30.43 to 520.8 ng/g lipid wt. The lowest and highest SCCP concentrations were found in *Mysis* and sculpin, respectively. Sum MCCP concentrations ranged from below the detection limit (*Mysis* and plankton) to as high as 2,250 ng/g lipid weight (sculpin). Note that the higher levels of sum MCCP concentrations observed in this study are consistent with the previous work of Houde et al. (2008). Because of the preliminary nature of the findings, and the difference in units relative to the other biological data, these results were not included in the statistical analysis.

As presented in Table 15a–c, limited data are available for the concentrations of chlorinated paraffins in the environment. SCCPs have been detected in all of the surface water samples ($n = 16$) from Lake Ontario and the St. Lawrence River and appear to be dominated by the $C_{10}$ and $C_{11}$ congeners (Table 15a). Total SCCP concentrations ranged from 0.0012 to 0.0384 μg/L; median and 95th percentile values were 0.012 and 0.0349 μg/L, respectively. MCCPs were also present in surface waters of Lake Ontario at substantially lower concentrations. SCCPs were present in all of the surficial sediment samples (Table 15b). Muir et al. (2001) provided concentrations for the SCCP congeners for eight samples, while only summary data were reported by Marvin et al. (2003). Total SCCP concentrations in Lake Ontario sediments range from 5.9 to 410 ng/g dwt; median and 95th percentile concentrations were 30 and 306.5 ng/g dwt, respectively. The residues of the $C_{12}$ congeners appear to be dominant based on limited data.

The results of mean concentrations reported for chlorinated paraffins in biological tissues are summarized in Table 15c. Note that the summary statistics represent concentrations for multiple trophic levels, ranging from invertebrates to small fish to predatory fish. Total SCCP concentrations ranged from 1.02 to 2,630 ng/g wwt, with median and 95th percentile concentrations of 24 and 624.4 ng/g wwt, respectively. In contrast to surface waters, both $C_{12}$ and $C_{13}$ congeners are predominant in biological tissues. As previously discussed, SCCP concentrations in fish vary considerably, depending on species and sampling location. The highest concentrations were observed in carp (2,630 ng/g wwt) as compared to lake trout (34–123 ng/g wwt). Total MCCP concentrations were lower, ranging from 0.05 to 109 ng/g wwt. However, as previously discussed, MCCP may be important in biota in certain lakes. The data for SCCPs in fish suggest that SCCP contamination appears to be widespread, but the concentrations are relatively low compared to other organic compounds (e.g., PCBs).

Risk assessments for exposure to the chlorinated paraffins (SCCP and MCCP) have been conducted by authorities in various geographies (EC 2008; EUC 2005c, 2008d; US EPA 1991, 1993). Although some risk assessments (United States, UK) have concluded that there is low ecological risk, others (Canada, Europe) have determined that these products pose a risk to aquatic life and should be phased out. Europe has recently completed risk assessments for both SCCP and MCCP (EUC 2005c, 2008d). Although no risks were indicated for certain uses, risk reduction

**Table 15** Summary statistics for chlorinated paraffins

| | *n* | Freq det (%) | Mean (μg/L) | SD (μg/L) | Median (μg/L) | 95th percentile (μg/L) | Min (μg/L) | Max (μg/L) |
|---|---|---|---|---|---|---|---|---|
| *A. Surface water* | | | | | | | | |
| C10 | 16 | 100 | 0.0045 | 0.0049 | 0.0032 | 0.0119 | 0.0001 | 0.0133 |
| C11 | 16 | 100 | 0.0059 | 0.0059 | 0.0050 | 0.0143 | 0.0004 | 0.0155 |
| C12 | 16 | 100 | 0.0032 | 0.0028 | 0.0018 | 0.0068 | 0.0007 | 0.0071 |
| C13 | 16 | 100 | 0.0014 | 0.0016 | 0.0009 | 0.0037 | 0.00004 | 0.0038 |
| Total SCCP | 16 | 100 | 0.0150 | 0.0143 | 0.0123 | 0.0349 | 0.0012 | 0.0384 |
| Total MCCP | 10 | 100 | 0.0000009 | | 0.0000009 | 0.0000009 | 0.0000009 | 0.0000009 |
| | *n* | Freq det (%) | Mean (ng/g dwt) | SD (ng/g dwt) | Median (ng/g dwt) | 95th percentile (ng/g dwt) | Min (ng/g dwt) | Max (ng/g dwt) |
| *B. Sediment* | | | | | | | | |
| C10 | 8 | 100 | 2.88 | 3.44 | 1.80 | 8.31 | 0.60 | 11.00 |
| C11 | 8 | 100 | 13.59 | 17.91 | 6.90 | 42.00 | 2.20 | 56.00 |
| C12 | 8 | 100 | 27.58 | 41.98 | 10.50 | 96.90 | 2.20 | 127.00 |
| C13 | 8 | 100 | 18.16 | 29.80 | 8.50 | 65.85 | 0.30 | 90.00 |
| Total SCCP | 38 | 100 | 69.24 | 97.28 | 30.00 | 306.50 | 5.90 | 410.00 |

**Table 15** (continued)

| | $n$ | Freq det (%) | Mean (ng/g wwt) | SD (ng/g wwt) | Median (ng/g wwt) | 95th percentile (ng/g wwt) | Min (ng/g wwt) | Max (ng/g wwt) |
|---|---|---|---|---|---|---|---|---|
| *C. Biota*[a] | | | | | | | | |
| C10 | 47 | 100 | 8.44 | 12.56 | 3.80 | 29.40 | 0.21 | 51.00 |
| C11 | 47 | 100 | 30.64 | 85.29 | 8.60 | 96.00 | 0.42 | 360.00 |
| C12 | 47 | 100 | 75.96 | 261.61 | 8.30 | 250.00 | 0.34 | 1,090.00 |
| C13 | 47 | 100 | 70.59 | 283.33 | 0.68 | 242.80 | 0.05 | 1,170.00 |
| Total SCCP | 47 | 100 | 185.55 | 630.74 | 24.00 | 624.40 | 1.02 | 2,630.00 |
| C14 | 44 | 61 | 7.30 | 11.39 | 2.60 | 28.90 | 0.05 | 38.00 |
| C15 | 44 | 61 | 10.55 | 21.30 | 0.48 | 57.50 | 0.05 | 64.00 |
| C16 | 44 | 36 | 2.87 | 6.06 | 0.05 | 15.75 | 0.05 | 19.00 |
| C17 | 44 | 14 | 0.25 | 0.48 | 0.05 | 1.28 | 0.05 | 1.60 |
| Total MCCP | 44 | 61 | 21.04 | 38.47 | 3.55 | 108.35 | 0.05 | 109.00 |

[a] The results represent concentrations reported for multiple trophic levels (invertebrates, small fish, large fish)

strategies have been implemented for their use in metal-working fluids, leather fat liquors, and in the production of polyvinylchloride. Europe has also concluded that SCCPs, but not MCCPs, meet criteria for persistence, bioaccumulation, and toxicity (PBT).

ENEVs and PNECs have been developed for chlorinated paraffins to be used in the calculation of risk quotients (exposure vs. effect). Comparison of the concentration distributions for SCCP and MCCP is summarized in Table 15a–c. These values, when compared with the ENEV and PNEC values shown in Table 2, suggest that all exposures are below the no effect values. However, Environment Canada (EC 2008) has recently concluded that, because of limitations in the available exposure and effects data, the absence of risk quotients greater than unity cannot be considered proof that these persistent and bioaccumulative substances do not cause ecological harm.

## 4 Summary

This review and statistical analysis was conducted to better understand the nature and significance of environmental exposures in the Great Lakes Basin and watershed to a variety of environmental contaminants. These contaminants of interest included current-use pesticides, pharmaceuticals, organic wastewater contaminants, alkylphenol ethoxylates, perfluorinated surfactants, flame retardants, and chlorinated paraffins. The available literature was critically reviewed and used to develop a database containing 19,611 residue values for 326 substances. In many papers, sampling locations were characterized as being downstream from municipal wastewater discharges, receiving waters for industrial facilities, areas susceptible to agricultural or urban contamination, or harbors and ports. To develop an initial assessment of their potential ecological significance, the contamination levels found were compared with currently available regulatory standards, guidelines, or criteria.

This review was prepared for the IJC multi-board work group, and served as background material for an expert consultation, held in March, 2009, in which the significance of the contaminants found was discussed. Moreover, the consultation attempted to identify and assess opportunities for strengthening future actions that will protect the Great Lakes.

Based on the findings and conclusions of the expert consultation, it is apparent that a wide variety of chemicals of emerging concern have been detected in environmental media (air, water, sediment, biota) from the Great Lakes Basin, although many are present at only trace levels. Although the presence of these contaminants raises concerns in the public and among the scientific community, the findings must be placed in context. Significant scientific interpretation is required to understand the extent to which these chemicals may pose a threat to the ecosystem and to human health. The ability to detect chemicals in environmental media greatly surpasses our ability to understand the implications of such findings. As advances in analytical technologies occur, it is probable that substances previously found to be non-detectable will be detected. However, their presence in environmental media should not be construed to mean that they are necessarily toxic or hazardous.

Current-use pesticides are tightly regulated and extensive efforts have been made to analyze for their presence in surface waters from the Great Lakes Basin. The concentrations found in surface waters for many of the pesticides are below current regulatory criteria. However, the concentrations of certain pesticides exceeded current criteria in 6–32% of the samples analyzed.

Detectable concentrations of pharmaceutical compounds were present in 34% of the surface water samples. Various prescription and non-prescription drugs were detected, most frequently at locations that were proximate to the point of discharge from wastewater treatment plants or agricultural operations. At present, there are no standards, guidelines, or criteria with which to compare these contaminant concentrations.

Concentrations of alkylphenol ethoxylates and their metabolites have been well studied. All surface water nonylphenol concentrations were below US ambient water quality criteria. However, the concentrations reported for some locations exceeded Canadian guidelines for water or sediment.

Only limited data were available for a wide variety of organic wastewater contaminants. Measured concentrations in Great Lakes waters were generally low. Where criteria exist for comparison, the concentrations found were generally below the associated regulatory standards. However, exceedences were noted for some classes of compounds, including phthalates and polycyclic aromatic hydrocarbons.

The highest environmental concentrations were reported in biota for a number of persistent, bioaccumulative, and toxic compounds (e.g., polybrominated diphenyl ethers, perfluorinated surfactants). Various stewardship as well as government risk assessment and risk management programs have been implemented over the past years for many of these compounds. Because risk management strategies for some of these contaminants have been implemented only recently, their impact on environmental concentrations, to date, remains unclear. Current evidence suggests that the concentrations of some brominated flame retardants are trending downward, while the concentrations of others appear to be increasing.

Regulatory criteria are not available for many of the chemicals of emerging concern that were detected in the Great Lakes Basin. When criteria do exist, it is important to recognize that they were developed based on the best available science at the time. As the science evolves, regulatory criteria must be reassessed in light of new findings (e.g., consideration of new endpoints and mechanisms of action). Further, there are significant scientific gaps in our ability to interpret environmental monitoring data, including the need for: (a) improving the understanding of the effects of mixtures, (b) information on use of, and the commercial life cycle of chemicals and products that contain them, (c) information on source contributions and exposure pathways, and (d) the need for thoughtful additional regulatory, environmental, and health criteria.

Discharges from wastewater treatment plants were identified as an important source of contaminants to surface waters in the Great Lakes Basin. Combined sewer overflows and agricultural operations were also found to be important contributors to concentrations in surface waters. Concentrations of many of the chemicals were generally the highest in the vicinity of these sources, decline with increasing distance from sources, and were generally low or non-detectable in the open waters of the Great Lakes.

**Acknowledgements** The authors wish to thank Sheridan Haack, Derek Muir, and Mike Murray for assistance with identifying relevant studies. The authors also thank John Struger for providing additional details on pesticide monitoring conducted in Canada.

## References

Advanced Industrial Science and Technology (AIST), National Institute of (2007) AIST Risk Assessment Document Series No. 4 Bisphenol A. Research Center for Chemical Risk Management, AIST Tsukuba West, Tsukuba, Ibaraki, JP

Andresen JA, Muir D, Ueno D, Darling C, Theobald N, Bester K (2007) Emerging pollutants in the North Sea in comparison to Lake Ontario, Canada. Environ Toxicol Chem 26:1081–1089

Balmer ME, Poiger T, Droz C, Romanin K, Bergqvist PA, Muller MD, Buser HR (2004) Occurrence of methyl triclosan, a transformation product of the bactericide triclosan, in fish from various lakes in Switzerland. Environ Sci Technol 38:390–395

Baronti C, Curini R, D'Ascenzo G, Dicorcia A, Gentili A, Samperi R (2000) Monitoring natural and synthetic estrogens at activated sludge sewage treatment plants and in a receiving river water. Environ Sci Technol 34:5059–5066

Batterman S, Chernyak S, Gwynn E, Cantonwine D, Jia C, Begnoche L, Hickey JP (2007) Trends of brominated diphenyl ethers in fresh and archived Great Lakes fish (1979–2005). Chemosphere 69:444–457

Bennett ER, Metcalf CD (1998) Distribution of alkylphenol compounds in Great Lakes sediments, United States and Canada. Environ Toxicol Chem 17:1230–1235

Bennett ER, Metcalfe CD (2000) Distribution of degradation products of alkylphenol ethoxylates near sewage treatment plants in the Lower Great Lakes, North America. Environ Toxicol Chem 19:784–792

Bennie DT, Sullivan CA, Lee HB, Peart TE, Maguire RJ (1997) Occurrence of alkylphenols and alkylphenol mono- and diethoxylates in natural waters of the Laurentian Great Lakes basin and the upper St. Lawrence River. Sci Total Environ 193:263–275

Boulanger B, Peck AM, Schnoor JL, Hornbuckle KC (2005) Mass budget of perfluorooctane surfactants in Lake Ontario. Environ Sci Technol 39:74–79

Boulanger B, Vargo J, Schnoor JL, Hornbuckle KC (2004) Detection of perfluorooctane surfactants in Great Lakes water. Environ Sci Technol 38:4064–4070

Boyd GR, Reemtsma H, Grimm DA, Mitra S (2003) Pharmaceuticals and personal care products (PPCPs) in surface and treated waters of Louisiana, USA and Ontario, Canada. Sci Total Environ 331:135–149

Calamari D, Jones K, Kannan K, Lecloux A, Olsson M, Thurman M, Zannetti P (2000) Monitoring as an indicator of persistence and long-range transport. In: Klečka GM (ed) Evaluation of persistence and long-range transport of organic chemicals in the environment. Society of Environmental Toxicology and Chemistry, Pensacola, FL, pp 205–244

Canadian Council of Ministers of the Environment (CCME) (2001) Canadian water quality guidelines for the protection of aquatic life: nonylphenol and its ethoxylates. Environment Canada Publication Number 12999. ISBN 10896997-34-1. Environment Canada, Ottawa, ON, Canada

CCME (2002) Canadian sediment quality guidelines for the protection of aquatic life: summary tables. Updated 2002. In: Canadian Council of Ministers of the Environment, Canadian environmental quality guidelines, 1999. Winnipeg, MB, Canada

CCME (2007) Canadian water quality guidelines for the protection of aquatic life: summary table. Update 7.1, December, 2007. In: Canadian Council of Ministers of the Environment, Canadian environmental quality guidelines, 1999. Winnipeg, MB, Canada
CCME (2008) Canadian water quality guidelines for the protection of aquatic life: chlorpyrifos. Updated 2008. In: Canadian Council of Ministers of the Environment, Canadian environmental quality guidelines, 1999. Winnipeg, MB, Canada
Canadian Environmental Protection Act (CEPA) (1997) Environmental assessments of priority substances under the Canadian Environmental Protection Act: guidance manual version 1.0. Chemical Evaluation Division, Commercial Chemicals Evaluation Branch, report EPS/2/CC/3E. Environment Canada, Ottawa, ON, Canada
Carlson DL, Basu I, Hites RA (2004) Annual variations of pesticide concentrations in Great Lakes precipitation. Environ Sci Technol 38:5290–5296
Carlson DL, Swackhamer DL (2006) Results from the U.S. Great Lakes fish monitoring program and effects of lake processes on bioaccumulative contaminant concentrations. J Great Lakes Res 32:370–385
Chernyak SM, Rice CP, Quintal RT, Begnoche LJ, Hickney JP, Vunyard BT (2005) Time trends (1983–1999) for organochlorines and polybrominated diphenyl ethers in rainbow smelt (*Osmerus mordax*) from Lakes Michigan, Huron, and Superior, USA. Environ Toxicol Chem 24:1632–1641
Christian MS, Parker RM, Hoberman AM, Diener RM, Api AM (1999) Developmental toxicity studies of four fragrances in rats. Toxicol Lett 111:169–174
Chu S, Haffner GD, Letcher RJ (2005) Simultaneous determination of tetrabromobisphenol A, tetrachlorobisphenol A, bisphenol A and other halogenated analogues in sediment and sludge by high performance liquid chromatography–electrospray tandem mass spectrometry. J Chromatogr A 1097:25–32
Chu S, Metcalfe CD (2007) Analysis of paroxetine, fluoxetine, and norfluoxetine in fish tissues using pressurized liquid extraction, mixed mode solid phase extraction cleanup, and liquid chromatography–tandem mass spectrometry. J Chromatogr A 1163:112–118
Cortes DR, Basu I, Sweet CW, Brice KA, Hoff RM, Hites RA (1998) Temporal trends in gas-phase concentrations of chlorinated pesticides measured at the shores of the Great Lakes. Environ Sci Technol 32:1920–1927
Daughton CG (2001) Pharmaceuticals in the environment: overarching issues and overview. In: Daughton CG, Jones-Lepp T (eds) Pharmaceuticals and personal care products in the environment: scientific and regulatory issues, American Chemical Society Symposium Series 791. Washington, DC, pp 2–38
Department for Environment, Food and Rural Affairs (DEFRA) (2004) Perfluorooctane sulphonate; risk reduction strategy and analysis of advantages and drawbacks. Final Report, August 2004. Norfolk, UK
Dsikowitzky L, Schwarzbauer J, Littke R (2002) Distribution of polycyclic musks in water and particulate matter of the Lippe River (Germany). Org Geochem 33:1747–1758
Environment Canada (EC) (2006a) Ecological screening assessment report on perfluorooctane sulfonate, its salts and its precursors that contain the $C_8F_{17}SO_2$, $C_8F_{17}SO_3$, or $C_8F_{17}SO_2N$ moiety. Environment Canada, Ottawa, ON, Canada
EC (2006b) Ecological screening assessment report on polybrominated diphenyl ethers (PBDEs). Environment Canada, Ottawa, ON, Canada
EC (2008) Follow-up report on a PSL1 assessment for which data were insufficient to conclude whether the substances were "toxic" to the environment and to human health: chlorinated paraffins. Environment Canada, Ottawa, ON, Canada
Environment Canada and Health Canada (EC and HC) (2001) Priority substances list assessment report: nonylphenol and its ethoxylates. Environment Canada and Health Canada, Ottawa, ON, Canada
EC and HC (2008) Screening assessment for the challenge: phenol, 4,4′-(1-methylethylidene)bis-(bisphenol A). Environment Canada and Health Canada, Ottawa, ON, Canada

European Commission (EUC) (2001) European Union risk assessment report: diphenyl ether, pentabromo derivative. Final Report. Office for Official Publications of the European Communities, Luxembourg

EUC (2002a) European Union risk assessment report: 4-nonylphenol (branched) and nonylphenol. Final Report. Office for Official Publications of the European Communities, Luxembourg

EUC (2002b) European Union risk assessment report: bis(pentabromophenyl) ether. Final report. Office for Official Publications of the European Communities, Luxembourg

EUC (2003a) Technical guidance document in support of Commission Directive 93/67/EEC on risk assessment for new notified substances and commission regulation (EC) No. 1488/94 on risk assessment for existing substances. Part II. ECSC-EC-EAEC. Brussels, Belgium

EUC (2003b) European Union risk assessment report: diphenyl ether, octabromo derivative. Final Report. Office for Official Publications of the European Communities, Luxembourg

EUC (2003c) Directive 2003/11/EC of the European Parliament and of the Council of 6 February 2003 amending for the 24th time Council directive 76/769 relating to restrictions on the marketing and use of certain dangerous substances and preparations (pentabromodiphenyl ether and octabromodiphenyl ether). Official Journal L 042

EUC (2003d) European Union risk assessment report: naphthalene. Final report. Office for Official Publications of the European Communities, Luxembourg

EUC (2004) European Union risk assessment report: 1,4-dichlorobenzene. Final report. Office for Official Publications of the European Communities, Luxembourg

EUC (2005a) 5-*tert*-Butyl 2,4,6-trinitro-*m*-xylene (musk xylene). Summary risk assessment report. European Chemicals Bureau, Ispra, Italy

EUC (2005b) European Union risk assessment report: 4′-*tert*-butyl-2′,6′-dimethyl-3′,5′-dinitroacetophenone (musk ketone). Final Report. Office for Official Publications of the European Communities, Luxembourg

EUC (2005c) European Union risk assessment report: alkanes, C14-17, chloro (MCCP): part 1 – environment. Final Report. Office for Official Publications of the European Communities, Luxembourg

EUC (2005d) European Union risk assessment report: tetrachloroethylene. Final report; part 1 – environment. Office for Official Publications of the European Communities, Luxembourg

EUC (2006a) Directive 2000/60/EC of the European Parliament and of the Council on environmental quality standards in the field of water policy and amending Directive 2000/60/EC. Official Journal COM(2006)397

EUC (2006b) European Union risk assessment report: phenol. Final report. Office for Official Publications of the European Communities, Luxembourg

EUC (2008a) Updated European Union risk assessment report – 4,4′-isopropylidenediphenol (bisphenol-A); CAS No: 80-05-7; EINECS No: 201-245-8. Final Report. Office for Official Publications of the European Communities, Luxembourg

EUC (2008b) European Union risk assessment report: 1-(5,6,7,8-tetrahydro-3,5,5,6,8,8-hexamethyl-2-naphthyl)ethan-1-one (AHTN). Final Report. Office for Official Publications of the European Communities, Luxembourg

EUC (2008c) European Union risk assessment report: 1,3,4,6,7,8-hexahydro-4,6,6,7,8,8-hexamethylcyclopenta-2-benzopyran (HHCB). Final Report. Office for Official Publications of the European Communities, Luxembourg

EUC (2008d) Updated risk assessment of alkanes, C10-13, chloro. Office for Official Publications of the European Communities, Luxembourg

EUC (2008e) European Union risk assessment report: anthracene. Final report. Office for Official Publications of the European Communities, Luxembourg

EUC (2008f) European Union risk assessment report: coal-tar pitch, high temperature. Final report. Office for Official Publications of the European Communities, Luxembourg

EUC (2008 g) European Union risk assessment report: hexabromocyclododecane. Final draft report. Office for Official Publications of the European Communities, Luxembourg

Fox GA (2006) Back to the future: rediscovering the requirement for monitoring in the Great Lakes Water Quality Agreement. In: Expert consultation on emerging issues of the Great Lakes in the 21st Century. Report of the Great Lakes Science Advisory Board to the International Joint Commission, Windsor, ONtario, Canada

Furdui VI, Stock NL, Ellis DA, Butt CM, Whittle DM, Crozier PW, Reiner EJ, Muir DCG, Mabury SA (2007) Spatial distribution of perfluoroalkyl contaminants in Lake Trout from the Great Lakes. Environ Sci Technol 41:1554–1559

Gatermann R, Biselli S, Huhnerfuss H, Rimkus GG, Hecker M, Karbe L (2002) Synthetic musks in the environment, part 1: species-dependent bioaccumulation of polycyclic and nitro musk fragrances in freshwater fish and mussels. Arch Environ Contam Toxicol 42:437–446

Gauthier LT, Hebert CE, Weseloh DVC, Letcher RJ (2007) Current-use flame retardants in the eggs of herring gulls (*Larus argentatus*) from the Laurentian Great Lakes. Environ Sci Technol 41:4561–4567

Gauthier LT, Herbert CE, Weseloh DVC, Letcher RJ (2008) Dramatic changes in the temporal trends of polybrominated diphenyl ethers (PBDEs) in herring gull eggs from the Laurentian Great Lakes: 1982–2006. Environ Sci Technol 42:1524–1530

Giesy JP, Kannan K (2001) Global distribution of perfluorooctane sulfonate in wildlife. Environ Sci Technol 35:1339–1342

Giesy JP, Naile JE, Khim JS, Jones PD, Newsted JL (2010) Aquatic toxicology of perfluorinated chemicals. Rev Environ Contam Toxicol 202:1–52

Gilliom RJ, Barbash JE, Crawford CG, Hamilton PA, Martin JD, Nakagaki N, Nowell LH, Scott JC, Stackelberg PE, Thelin TP, Wolock DM (2006) The quality of our Nation's waters: pesticides in the Nation's streams and ground water, 1992–2001. U.S. Geological Survey Circular 1291, Reston, VA

Gouin T, Harner T, Daly GL, Wania F, Mackay D, Jones KC (2005) Variability of concentrations of polybrominated diphenyl ethers and polychlorinated biphenyls in air: implications for monitoring, modeling and control. Atmos Environ 39:151–166

Hao C, Lissemore L, Nguyen B, Kleywegt S, Yang P, Solomon K (2006) Determination of pharmaceuticals in environmental waters by liquid chromatography/electrospray ionization/tandem mass spectrometry. Anal Bioanal Chem 384:505–513

Hekster FM, de Voogt P, Pijnenburg AMCM, Laane RWPM (2002) Perfluoroalkylated substances; aquatic environmental assessment. RIKZ Report 2002.043, July 1, 2002. Den Haag, Netherlands

Hoh E, Hites RA (2005) Brominated flame retardants in the atmosphere of the east-central United States. Environ Sci Technol 39:7794–7802

Hoh E, Zhu L, Hites RA (2005) Novel flame retardants, 1,2-bis(2,4,6-tribromophenoxy)-ethane and 2,3,4,5,6-pentabromoethylbenzene, in United States' environmental samples. Environ Sci Technol 39:2472–2477

Hoh E, Zhu LY, Hites RA (2006) Dechlorane Plus, a chlorinated flame retardant, in the Great Lakes. Environ Sci Technol 40:1184–1189

Houde M, Muir DCG, Tomy GT, Whittle DM, Teixeira C, Moore S (2008) Bioaccumulation and trophic magnification of short- and medium-chain chlorinated paraffins in food webs from Lake Ontario and Lake Michigan. Environ Sci Technol 42:3893–3899

Hua WY, Bennett ER, Letcher RJ (2005) Triclosan in waste and surface waters from the upper Detroit River by liquid chromatography–electrospray–tandem quadrupole mass spectrometry. Environ Int 31:621–630

Hua WY, Bennett ER, Letcher RJ (2006a) Ozone treatment and the depletion of detectable pharmaceuticals and atrazine herbicide in drinking water sourced from the upper Detroit River, ON, Canada. Water Res 40:2259–2266

Hua WY, Bennett ER, Maio X, Metcalfe CD, Letcher RJ (2006b) Seasonality effects on pharmaceuticals and *s*-triazine herbicides in wastewater effluent and surface water from the Canadian side of the upper Detroit River. Environ Toxicol Chem 25:2356–2365

International Joint Commission (IJC) (1987) Revised Great Lakes water quality agreement of 1978 – agreement with annexes in terms of reference, between the United States and Canada signed at Ottawa October 16, 1983 as amended by protocol signed November 18, 1987. International Joint Commission Canada and United States, Windsor, ON, Canada

James R, Hites RA (1999) Chlorothalonil and dacthal in Great Lakes air and precipitation samples. J Great Lakes Res 25:406–411

Jones-Lepp T (2006) Chemical markers of human waste contamination: analysis of urobilin and pharmaceuticals in source waters. J Environ Monit 8:472–478

Kannan K, Kober JL, Kang Y, Masunaga S (2001) Polychlorinated naphthalenes, biphenyls, dibenzo-*p*-dioxins, and dibenzofurans as well as polycyclic aromatic hydrocarbons and alkylphenols in sediment from the Detroit and Rouge Rivers, Michigan, USA. Environ Toxicol Chem 20:1878–1889

Kannan K, Corsolini S, Falandysz J, Oehme G, Focardi S, Giesy JP (2002) Perfluorooctanesulfonate and related fluorinated hydrocarbons in marine mammals, fishes, and birds from coasts of the Baltic and the Mediterranean Seas. Environ Sci Technol 36:3210–3216

Kannan K, Keith TL, Naylor CG, Staples CA, Giesy JP (2003) Nonylphenol and nonylphenol ethoxylates in fish, sediment and water from the Kalamazoo River, Michigan. Arch Environ Contam Toxicol 4:77–82

Kannan K, Yun SH, Evans TJ (2005a) Chlorinated, brominated, and perfluorinated contaminants in livers of polar bears from Alaska. Environ Sci Technol 39:9057–9063

Kannan K, Tao L, Sinclair E, Pastva SD, Jude DJ, Giesy JP (2005b) Perfluorinated compounds in aquatic organisms at various trophic levels in a Great Lakes food chain. Arch Environ Contam Toxicol 48:559–566

Keith TL, Snyder SA, Naylor CG, Staples CA, Summer C, Kannan K, Giesy JP (2001) Identification and quantitation of nonylphenol ethoxylates and nonylphenol in fish tissues from Michigan. Environ Sci Technol 35:10–13

Klečka GM, Zabik J, Woodburn K, Naylor CG, Staples CA, Huntsman BE (2007) Exposure analysis of C8- and C9-alkylphenol, alkylphenol ethoxylates and their metabolites in surface water systems within the United States. Human Ecol Risk Assess 13:792–822

Klečka GM, Staples CA, Clark KE, van der Hoeven N, Thomas DE, Hentges SG (2009) Exposure analysis of bisphenol A in surface water systems in North America and Europe. Environ Sci Technol 43:6145–6150

Kolpin DW, Furlong ET, Meyer MT, Thurman EM, Zaugg SD, Barber LB, Buxton HT (2002) Pharmaceuticals, hormones, and other organic wastewater contaminants in U.S. streams, 1999–2000: a national reconnaissance. Environ Sci Technol 36:1202–1211

Lehman-McKeeman LD, Caudill D, Vassallo JD, Pearce RE, Madan A, Parkinson A (1999) Effects of musk xylene and musk ketone on rat hepatic cytochrome P450 enzymes. Toxicol Lett 111:105–115

Lindstrom A, Buerge IJ, Poiger T, Bergqvist PA, Muller MD, Buser HR (2002) Occurrence and environmental behavior of the bactericide triclosan and its methyl derivative in surface waters and in wastewater. Environ Sci Technol 36:2322–2329

Lissemore L, Hao C, Yang P, Sibley PK, Mabury S, Solomon K (2006) An exposure assessment for selected pharmaceuticals within a watershed in Southern Ontario. Chemosphere 64: 717–729

Lopes TJ, Furlong ET (2001) Occurrence and potential adverse effects of semi volatile organic compounds in streambed sediment, United States, 1992–1995. Environ Toxicol Chem 20: 727–737

Luross JM, Alaee M, Sergeant DB, Cannon CM, Whittle DM, Solomon KR, Muir DCG (2002) Spatial distribution of polybrominated diphenyl ethers and polybrominated biphenyls in lake trout from the Laurentian Great Lakes. Chemosphere 46:665–672

Manchester-Neesvig JB, Valters K, Sonzogni WC (2001) Comparison of polybrominated diphenyl ethers (PBDEs) and polychlorinated biphenyls (PCBs) in Lake Michigan salmonids. Environ Sci Technol 35:1072–1077

Martin JW, Whittle DM, Muir DCG, Mabury SA (2004) Perfluoroalkyl contaminants in a food web from Lake Ontario. Environ Sci Technol 38:5379–5385

Martin JW, Ellis DA, Mabury SA, Hurley MD, Wallington TJ (2006) Atmospheric chemistry of perfluoroalkanesulfonamides: kinetic and product studies of the OH radical and Cl atom initiated oxidation of *N*-ethyl perfluorobutanesulfonamide. Environ Sci Technol 40:864–872

Marvin CH, Painter S, Tomy GT, Stern GA, Braekevelt E, Muir DCG (2003) Spatial and temporal trends in short-chain chlorinated paraffins in Lake Ontario sediments. Environ Sci Technol 37:4561–4568

Marvin CH, Tomy GT, Alaee M, Macinnis G (2006) Distribution of hexabromocyclododecane in Detroit River suspended sediments. Chemosphere 64:268–275

Mayer T, Bennie D, Rosa F, Rekas G, Palabrica V, Schachtschneider J (2007) Occurrence of alkylphenolic substances in a Great Lakes coastal marsh, Cootes Paradise, ON, Canada. Environ Pollut 147:683–900

McDowell DC, Metcalfe CD (2001) Phthalate esters in sediments near a sewage treatment plant outflow in Hamilton Harbour, Ontario: SFE extraction and environmental distribution. J Great Lakes Res 27:3–9

Melcer H, Klečka G, Monteith H, Staples C (2007) Wastewater Treatment of Alkylphenols and their Ethoxylates; a State of the Science Review. Report for the Water Environment Federation, Alexandria, VA

Metcalfe CD, Miao XS, Koenig BG, Struger J (2003) Distribution of acidic and neutral drugs in surface waters near sewage treatment plants in the lower Great Lakes, Canada. Environ Toxicol Chem 22:2881–2889

Miller SM, Sweet CW, DePinto JV, Hornbuckle KC (2000) Atrazine and nutrients in precipitations: results from the Lake Michigan mass balance study. Environ Sci Technol 34:55–61

Monteiro SC, Boxall BA (2010) Occurrence and fate of human pharmaceuticals in the environment. Rev Environ Contam Toxicol 202:53–154

Moore S, Vromet L, Rondeau B (2004) Comparison of meta-stable atom bombardment and electron capture negative ionization for the analysis of polychloroalkanes. Chemosphere 54:453–459

Muir D, Alaee M, Scott B (2006) Identifying potential and emerging chemical contaminants in the Great Lakes. In: Expert consultation on emerging issues of the Great Lakes in the 21st Century. Report of the Great Lakes Science Advisory Board to the International Joint Commission, Windsor, ON, Canada

Muir D, Bennie D, Teixeria C, Fisk A, Tomy G, Stern G, Whittle M (2001) Short chain chlorinated paraffins: are they persistent and bioaccumulative? In: Lipnick RL, Jannson B, Mackay D (eds) Persistent, bioaccumulative, and toxic chemicals II. Assessment and new chemicals. ACS Symposium Series 773, American Chemical Society, Washington, DC

Norstrom RJ, Simon M, Moisey J, Wakeford B, Weseloh DVC (2002) Geographical distribution (2000) and temporal trends (1981–2000) of brominated diphenyl ethers in Great Lakes herring gull eggs. Environ Sci Technol 36:4786–4789

Organization for Economic Cooperation and Development (OECD) (2002) Cooperation on existing chemicals: hazard assessment of perfluorooctane sulfonate and its salts. Environment Directorate, Joint Meeting of the Chemicals Committee and the Working Party on Chemicals, Pesticides and Biotechnology, report ENV/JM/RD(2002)17/Final. Paris, France

O'Toole S, Metcalfe C (2006) Synthetic musks in fish from urbanized areas of the lower Great Lakes, Canada. J Great Lakes Res 32:361–369

Peck AM, Hornbuckle KC (2004) Synthetic musk fragrances in Lake Michigan. Environ Sci Technol 38:367–372

Peck AM, Hornbuckle KC (2006) Synthetic musk fragrances in urban and rural air of Iowa and the Great Lakes. Atmos Environ 40:6101–6111

Peck AM, Linebaugh EK, Hornbuckle KC (2006) Synthetic musk fragrances in Lake Erie and Lake Ontario sediment cores. Environ Sci Technol 40:5629–5635

Purdom CE, Hardiman PA, Bye VJ, Eno NC, Tyler CR, Sumpter JP (1994) Estrogenic effects of effluents from sewage treatment works. Chem Ecol 8:275–285

Qiu X, Marvin CH, Hites RA (2007) Dechlorane plus and other flame retardants in a sediment core from Lake Ontario. Environ Sci Technol 41:6014–6019

Rice CP, Chernyak SM, Begnoche L, Quintal R, Hickey J (2002) Comparisons of PBDE composition and concentration in fish collected from the Detroit River, MI and Des Plaines River, IL. Chemosphere 49:731–737

Rice CP, Schmitz-Afonso I, Loyo-Rosales JE, Link E, Thoma R, Fay L, Altfater D, Camp MJ (2003) Alkylphenol and alkylphenol ethoxylates in carp, water, and sediment from the Cuyahoga River, Ohio. Environ Sci Technol 37:3747–3754

Ricking M, Schwarzbauer J, Hellou J, Svenson A, Zitko V (2003) Polycyclic aromatic musk compounds in sewage treatment plant effluents of Canada and Sweden – first results. Mar Pollut Bull 46:410–417

Rimkus G, Rimkus B, Wolf M (1994) Nitro musks in human adipose tissue and breast milk. Chemosphere 28:421–432

Rimkus GG, Butte W, Geyer HJ (1997) Critical considerations on the analysis and bioaccumulation of musk xylene and other synthetic nitro musks in fish. Chemosphere 35:1497–1507

Rimkus GG (1999) Polycyclic musk fragrances in the aquatic environment. Toxicol Lett 111: 37–56

Routledge EJ, Sheahan D, Desbrow C, Sumpter JP (1998) Identification of estrogenic chemicals in STW effluent. 2. In vivo responses in trout and roach. Environ Sci Technol 32: 1559–1565

Sabik H, Gagne F, Blaise C, Marcogliese DJ, Jeannot R (2003) Occurrence of alkylphenol polyethoxylates in the St. Lawrence River and their bioconcentration by mussels (*Elliptio complanata*). Chemosphere 51:349–356

Sandstrom MW, Koplin DW, Thurman M, Zaugg SD (2005) Widespread detection of *N,N*-diethyl-*m*-toluamide in U.S. streams: comparison with concentrations of pesticides, personal care products, and other organic wastewater compounds. Environ Toxicol Chem 24:1029–1034

Schultz MM, Barofsky DF, Field JA (2003) Fluorinated alkyl surfactants. Environ Eng Sci 20: 487–501

Scott BF, Moody CA, Spencer C, Small JM, Muir DCG, Mabury SA (2006) Analysis for perfluorocarboxylic acids/anions in surface waters and precipitation using GC–MS and analysis of PFOA from large-volume samples. Environ Sci Technol 40:6405–6410

Servos MR, Maguire RJ, Bennie DT, Lee HB, Cureton PM, Davidson N, Sutcliff R, Rawn DFK (2003) An ecological risk assessment of nonylphenol and its ethoxylates in the aquatic environment. Hum Ecol Risk Assess 9:569–587.

Servos MR, Smith M, McInnis R, Burnison BK, Lee HB, Seto P, Backus S (2007) The presence of selected pharmaceuticals and the antimicrobial triclosan in drinking water in Ontario, Canada. Water Qual Res J Can 42:130–137

Shen L, Wania F, Lei YD, Teixeira CD, Muir DCG, Xiao H (2006) Polychlorinated biphenyls and polybrominated diphenyl ethers in the North American atmosphere. Environ Pollut 144: 434–444

Simonich SL, Begley WM, Debaere G, Eckhoff WS (2000) Trace analysis of fragrance materials in wastewater and treated wastewater. Environ Sci Technol 34:959–965

Sinclair E, Mayack DT, Roblee K, Yamashita N, Kannan K (2006) Occurrence of perfluoroalkyl surfactants in water, fish, and birds from New York State. Arch Environ Contam Toxicol 50:398–410

Song W, Ford JC, Li A, Mills WJ, Rockne KJ, Buckley DR (2004) Polybrominated diphenyl ethers in the sediments of the Great Lakes. 1. Lake Superior. Environ Sci Technol 38:3286–3293

Song W, Ford JC, Li A, Sturchio NC, Rockne KJ, Buckley DR, Mills WJ (2005a) Polybrominated diphenyl ethers in the sediments of the Great Lakes. 3. Lakes Ontario and Erie. Environ Sci Technol 39:5600–5605

Song W, Li A, Ford JC, Sturchio NC, Rockne KJ, Buckley DR, Mills WJ (2005b) Polybrominated diphenyl ethers in the sediments of the Great Lakes. 2. Lakes Michigan and Huron. Environ Sci Technol 39:3474–3479

Strandberg B, Dodder NG, Basu I, Hites RA (2001) Concentrations and spatial variations of polybrominated diphenyl ethers and other organohalogen compounds in Great Lakes air. Environ Sci Technol 35:1078–1083

Streets S, Henderson S, Stoner A, Carlson D, Simcik M, Swackhamer D (2006) Partitioning and bioaccumulation of PBDEs and PCBs in Lake Michigan. Environ Sci Technol 40:7263–7269

Struger J, Fletcher T (2007) Occurrence of lawn care and agricultural pesticides in the Don and Humber River Watersheds (1998–2002). J Great Lakes Res 33:887–905

Struger J, L'Italien S, Sverko E (2004) In-use pesticide concentrations in surface waters of the Laurentian Great Lakes (1994–2000). J Great Lakes Res 30:435–450

Struger J, Sverko E, Grabuski J, Fletcher T, Marvin C (2007) Occurrence and fate of methoprene compounds in urban areas of Southern Ontario, Canada. Bull Environ Contam Toxicol 79: 168–171

Struger J, Thompson D, Staznik B, Martin P, McDaniel T, Marvin C (2008) Occurrence of glyphosate in surface waters of southern Ontario. Bull Environ Contam Toxicol 80:378–384

Sun P, Backus S, Blanchard P, Hites RA (2006a) Temporal and spatial trends of organochlorine pesticides in Great Lakes precipitation. Environ Sci Technol 40:2135–2141

Sun P, Blanchard P, Brice K, Hites RA (2006b) Atmospheric organochlorine pesticide concentrations near the Great Lakes: temporal and spatial trends. Environ Sci Technol 40:6587–6593

Sverko E, Tomy GT, Marvin CH, Zaruk D, Reiner E, Helm PA, Hill B, McCarry BE (2008) Dechlorane plus levels in sediment of the lower Great Lakes. Environ Sci Technol 42:361–366

Tertuliana JS, Alvarez DA, Furlong ET, Meyer MT, Zaugg SD, Koltun GF (2008) Occurrence of organic wastewater compounds in the Tinkers Creek watershed and two other tributaries to the Cuyahoga river, Northeast Ohio. USGS Report 2008-5173, Reston, VA

Tomy GT, Budakowski W, Halldorson T, Whittle DM, Keir MJ, Marvin C, MacInnis G, Alaee M (2004) Biomagnification of alpha- and gamma-hexabromocyclododecane isomers in a Lake Ontario food web. Environ Sci Technol 38:2298–2303

Tomy GT, Pleskach K, Ismail N, Whittle DM, Helm PA, Sverko E, Zaruk D, Marvin CH (2007) Isomers of Dechlorane plus in Lake Winnipeg and Lake Ontario food webs. Environ Sci Technol 41:2249–2254

Tuduri L, Harner T, Blanchard P, Li YF, Poissant L, Waite DT, Murphy C, Belzer W (2006) A review of currently used pesticides (CUPs) in Canadian air and precipitation. Part 2: regional information and perspectives. Atmos Environ 40:1579–1589

United Kingdom Environment Agency (UK EA) (2005) Environmental risk evaluation report: 4-*tert*-octylphenol. London, UK

United States Environmental Protection Agency (US EPA) (1991) RM1 decision package: chlorinated paraffins – environmental risk assessment. US Environmental Protection Agency, Washington, DC

United States Environmental Protection Agency (US EPA) (1993) RM2 exit briefing on chlorinated paraffins and olefins. US Environmental Protection Agency, Washington, DC

United States Environmental Protection Agency (US EPA) (1996) RM-1 document for *para*-nonylphenol. US Environmental Protection Agency, Washington, DC

United States Environmental Protection Agency (US EPA) (1998) Reregistration eligibility decision (RED) for DEET. EPA 738-R-98-010. US Environmental Protection Agency, Washington, DC

United States Environmental Protection Agency (US EPA) (1999) National recommended water quality criteria. Report 822-Z-99-001. U.S. Environmental Protection Agency, Washington, DC

U.S. Environmental Protection Agency (US EPA) (2005) Aquatic life ambient water quality criteria – nonylphenol. Report 822-R-05-005. U.S. Environmental Protection Agency, Washington, DC

U.S. Environmental Protection Agency (US EPA) (2006) National recommended water quality criteria. Office of Water (4304T). U.S. Environmental Protection Agency, Washington, DC

Valters K, Li H, Alaee M, D'Sa I, Marsh G, Bergman A, Letcher RJ (2005) Polybrominated diphenyl ethers and hydroxylated and methoxylated brominated and chlorinated analogues in the plasma of fish from the Detroit River. Environ Sci Technol 39:5612–5619

Venier M, Hites RA (2008) Flame retardants in the atmosphere near the Great Lakes. Environ Sci Technol 42:4745–4751

Walker JD (2006) Can QSARs identify emerging pollutants that will threaten the Great Lakes ecosystem? In: Expert consultation on emerging issues of the Great Lakes in the 21st century. Report of the Great Lakes Science Advisory Board to the International Joint Commission, Windsor, ON, Canada

Walraven N, Laane RWPM (2008) Assessing the discharge of pharmaceuticals along the Dutch coast of the North Sea. Rev Environ Contam Toxicol 199:1–18

Winkler M, Kopf G, Hauptvogel C, Neu T (1998) Fate of artificial musk fragrances associated with suspended particulate matter (SPM) from the River Elbe (Germany) in comparison to other organic contaminants. Chemosphere 37:1139–1156

Yamagishi T, Miyazaki T, Horii S, Kaneko S (1981) Identification of musk xylene and musk ketone in fresh-water fish collected from the Tama River, Tokyo. Bull Environ Contam Toxicol 26: 656–662

Zhu LY, Hites RA (2004) Temporal trends and spatial distributions of brominated flame retardants in archived fishes from the Great Lakes. Environ Sci Technol 38:2779–2784

# The Elderly as a Sensitive Population in Environmental Exposures: Making the Case

John F. Risher, G. Daniel Todd, Dean Meyer, and Christie L. Zunker

## Contents



## 1 Introduction

The population of the United States is aging. A demographic shift of the average age in the United States is occurring rapidly and unavoidably. According to estimates, 20% of all Americans will be 65 or older by the year 2030 (CDC 2004), which is a serious concern for physicians and medical science (Geokas et al. 1990). Further,

J.F. Risher (✉)
Agency for Toxic Substances and Disease Registry, Division of Toxicology (F-32), Toxicology Information Branch, 1600 Clifton Road, Atlanta, GA 30333, USA
e-mail: jrisher@cdc.gov

D.M. Whitacre (ed.), *Reviews of Environmental Contamination and Toxicology*,
Reviews of Environmental Contamination and Toxicology 207,
DOI 10.1007/978-1-4419-6406-9_2, 

the average 75-year-old has three chronic medical conditions and uses five prescription drugs (CDC 2004; Qato et al. 2008). It is now recognized that infants and small children are not just miniature adults. Children have many unique anatomic, developmental, physiologic, immunologic, and psychological considerations (Milla 2002; WHO 2005; Williams et al. 2006). They have special nutritional and medical needs because of the immature state of their physiological and anatomical development. Similarly, the elderly are not just older adults with the same needs as healthy young adults. The US Centers for Disease Control and Prevention (CDC) recognizes this difference and states that older adults have unique challenges and different medical needs than younger adults (CDC 2004).

Despite the fact that the elderly have been viewed as a population at particular risk to environmental chemicals for some time due to their increased sensitivity (Birnbaum 1991), there is a paucity of published data that address the effects of environmental substances in an elderly population. Nonetheless, the physiologic processes that metabolize and eliminate xenobiotics, including pharmaceuticals, are known to be significantly compromised as a result of the normal aging process in humans. The purpose of this chapter is to present what is known about the aging process with respect to the body's ability to deal with exposure to xenobiotics, including environmental chemicals. It is the further intent of the authors to present the reasons that these age-related changes might make the elderly more susceptible to the adverse effects of environmental substances. And while all of the physiologic systems of the body undergo degradation during the aging process, this chapter addresses only those systems and those age-related effects that may increase the risk posed by environmental toxicants.

## 2 What Is Aging and What Makes the Elderly Particularly Susceptible to Environmental Chemicals?

Aging has been defined in several ways, all of which include a gradual deterioration in body function and the capacity to respond to environmental stresses. Partridge and Mangel (1999) defined aging as a progressive functional decline or a gradual deterioration of physiological function with age. Another definition is that aging is a time of loss of homeostatic reserve and of reduced adaptability to metabolic perturbation (Mooradian 1992). Harman (1981) described aging as the progressive accumulation of changes with time that are associated with, or responsible for, the ever-increasing susceptibility to disease and death, which accompanies advancing age. And Wagner et al. (2008) describe aging as a complex process that involves every cell and organ in the body and that leads to the deterioration of many body functions over the life span of the individual. Put another way, aging is associated with a wide range of physiologic changes that limit our normal functions and render us more susceptible to a number of diseases.

Among the elderly, functional disability occurs faster and takes longer to remediate (Oskvig 1999). Although many millions of new cells normally are produced in most tissues throughout life, heart muscle, skeletal muscle, and most brain cells are not replaced to a biologically relevant extent, if at all. Many of the common effects of aging are related to both a decrease in number of cells and to dysfunction in some cells that remain.

Advanced age in most species is associated with impaired adaptive and homeostatic mechanisms, leading to susceptibility to environmental or internal stress with resultant increases in rates of disease and death (Grimley Evans 2000). The ability of the body to respond to physiologic challenge imposed by potentially toxic substances in the environment is dependent upon the health of the organ systems that eliminate those substances from the body. Age-related changes in sensitivity to environmental chemicals can result from alterations in either toxicokinetic (what the body does to the chemical) or toxicodynamic (what the chemical does to the body) processes (Birnbaum 1991). Pathologic states that compromise the function of any of the organ systems cause a decrease in the body's ability to protect itself from the adverse effects of exposure to those contaminants. From the age of 30, a number of our physiologic systems begin to decline; and as we age, our homeostatic reserves decline, adversely affecting the ability to respond to environmental change or toxic insult (NRC 1987).

## *2.1 Aging and the Kidney*

With advancing age, the kidneys become less effective in performing their task of filtering the blood. The kidneys, which are organs critical for waste elimination, maintenance of electrolyte and water balance, and the pH of the blood, progressively shrink in size with aging. The effects of aging on the kidneys include a progressive deterioration of both renal structure and function (Anderson and Brenner 1986; Baylis 2005; Lindeman and Preuss 1994). Vital roles of the kidney, such as the active elimination of hydrogen ions to maintain blood pH, are compromised. Thus, the ability of the kidneys to restore physiological pH in times of serum acidosis is compromised in the elderly (Beers and Berkow 2000).

Gourtsoyiannis et al. (1990) studied 360 adult men and women with ages ranging from 20 to 80 years and found an approximate 10% decrease in renal parenchyma per decade of life. A decrease was observed in both kidneys and in both sexes, with a higher rate of decrease occurring between the sixth and seventh decades (Gourtsoyiannis et al. 1990).

At age 30, the adult kidney weighs approximately 150–200 g (Hazzard et al. 1999). By the age of 90, the average weight has fallen to 110–150 g, or a loss of 20–30% in organ weight, with a 40% decrease in volume during that period. Most of the tissue loss is from the renal cortex (Tauchi et al. 1971) and is accompanied by glomerular and tubular loss. Between the ages of 30 and 80, the total number of glomeruli decreases. By age 70, there is a 30–50% loss of functioning glomeruli due to age alone (Lindeman 1993).

An increase in the prevalence of glomerulosclerosis occurs with advancing age (Kaplan et al. 1975; Kasiske 1987; Neugarten et al. 1999). This increase is independent of gender (Neugarten et al. 1999). Age-associated glomerulosclerosis has been described as being very similar to glomerular sclerosis induced by other pathological processes (Lopez-Novoa 2008). Kaplan et al. (1975) reported that the incidence of glomerulosclerosis is less than 10% in individuals under 40 years of age for their study population, but increases in those over 40 and accelerates in those over 50 in a variable fashion. Kappel and Olsen (1980) examined 123 kidneys, 54 intended for transplantation and 69 from autopsy, and reported that the relative number of sclerotic, obsolescent glomeruli was highly dependent on age. That number was very small (0–1%) until the age of 40, increasing thereafter until it reached values of about 30% in persons more than 80 years old. The glomerular basement membrane undergoes progressive folding and thickening and then condenses into hyaline material, with the subsequent collapse of the glomerular tuft. Free anastomoses are then formed among the reduced number of capillary loops which further complicates renal function (Lopez-Novoa 2008).

Functionally, glomerular filtration and renal blood flow rates decline in a linear fashion after the age of 30 (Anderson and Brenner 1986). In octogenarians, these values have been reduced to one-half to two-thirds of the values measured in young adults. The decrease in renal blood flow and glomerular filtration rate (GFR) reduces the body's ability to properly filter the blood and eliminate biological wastes. This decrease in GFR can also serve to potentiate the biologic effects of some xenobiotics by increasing the residence time of (water-soluble) metabolites of environmental chemicals in the body. Such an increased availability in the blood can result in greater distribution to potential target tissues. In the case of some substances, such as ionic forms of heavy metals, the half-life of these substances would no doubt increase.

In addition, polyuria, nocturia, increased frequency of urination, dysuria, urine retention in the bladder, and hematuria may occur, as may acute and chronic kidney inflammation and renal calculi (Costa-Bauza et al. 2007; Graugaard-Jensen et al. 2008; Homma et al. 1994). Subsequent alterations in water balance and decreased sense of thirst increase the risk of dehydration. In aging men, prostate problems become more common. Benign prostatic hyperplasia (BPH) is clinically evident in 50% of men by age 50 and in 80% by age 80 (Beers and Berkow 2000). Androgens, particularly dihydrotestosterone, appear to play a major role.

Because the prostate encircles part of the urethra, enlargement of this gland can cause urine retention and difficulty in urination. This can also create back pressure on the kidneys and result in further problems in the kidney itself. Hyperplasia of the prostate, with subsequent increase in the fibromuscular stroma, results in a narrowing of the urethral lumen as it traverses the prostate. This narrowing creates bladder outlet obstruction. In addition, prostatic smooth muscle tone, mediated through α-adrenergic receptors, creates further bladder outlet obstruction (Beers and Berkow 2000).

Advancing age is typically accompanied by a decrease in the size of the urinary bladder, resulting in a decrease in bladder volume (Chutka et al. 1996). This contributes to polyuria and nocturia. In addition, many elderly individuals experience premature contractions of the muscular detrusor, even when the urine content in the bladder is low. Urinary tract infections, enlargement of the prostate, pelvic floor dysfunction, fecal impactions, and immobility also contribute to urinary incontinence among the elderly (Chutka et al. 1996).

Because of the age-associated changes in urinary system function, the elderly are significantly more vulnerable to injury from a variety of environmental and pharmacological agents (Hazzard et al. 1999). The reduction in GFR can lead to a symptomatic retention of nitrogenous wastes in aged individuals. Reduced GFR also results in a decreased renal clearance of prescription drugs, such as digoxin, and many environmental substances. This provides for a longer residence time in the body and the possible accumulation through subsequent exposures (Lindeman et al. 1985).

Compromised renal clearance, as indicated by an increase in the elimination half-life of a number of pharmacologic preparations, has been reported to increase with advancing age (Hazzard et al. 1999). For example, Luvox has been reported to be cleared more slowly in elderly persons (ages 66–73) than in younger adults (ages 19–35), with a difference in the elimination half-time of approximately twice that of younger adults. Mean plasma concentrations of the parent compound were approximately 40% higher in the elderly group (PDR 2007).

Although there are no studies to show that this also occurs with environmental toxicants, there is no reason to believe that their elimination would not be similarly affected. Thus, the effects of exposure of the elderly to environmental toxicants may be more severe than they would be in a healthy, young adult. Further, concomitant prescription drug intake to treat various age-related conditions could result in a greater negative impact than would individual exposures to environmental or pharmaceutical substances. Some pharmaceuticals and environmental toxicants known to adversely affect the kidneys are shown in Table 1.

**Table 1** Some known renal toxicants and their effects in laboratory mammals and/or humans

| Toxicant | Use/Source | Effect | Reference |
|---|---|---|---|
| Acetaminophen | Analgesic, anti-pyretic, anti-inflammatory | Proximal tubular necrosis; acute renal failure (large doses) | Emeigh Hart et al. (1994, 1996) |
| Ibuprofen, naproxen, and indomethacin | Analgesic, anti-inflammatory agent | Nephropathy characterized by papillary necrosis with chronic interstitial nephritis | Murray and Brater (1993) |
| Mercury | Thermometers, barometers, many industrial/other uses | Increased kidney weight; inflammation; slight histopathologic changes in cortex; tubular degeneration/atrophy; proximal tubular necrosis; fibrosis | Dieter et al. (1992); Hultman and Enestrom (1992); Jonker et al. (1993); NTP (1993) |

**Table 1** (continued)

| Toxicant | Use/Source | Effect | Reference |
|---|---|---|---|
| Cadmium | Ni–Cd batteries; pigments for plastics, ceramics, and glasses; stabilizers for PVC; coatings on steel and non-ferrous metals; photography; photocopying; lubricants | Increased proteinuria; proximal tubular lesions/necrosis/ dysfunction; chronic interstitial nephritis; glomerular and interstitial fibrosis | Cha (1987); Fingerle et al. (1982); Kotsonis and Klaassen (1978); Prigge (1978); Shiwin et al. (1990); Stowe et al. (1972) |
| Haloalkenes and halobenzenes | Solvents/degreasers; synthesis of various organics; synthesis of chlorinated solvents; extraction of rubber and oils; refrigerants; space deodorants; fumigants for moth/mold/mildew control | Increased kidney weight; hemoglobinuria; proximal tubular necrosis; renal tubular degeneration; renal failure | Hayes et al. (1987); Henck et al. (1979); Lee et al. (1977); Maltoni et al. (1985); McCauley et al. (1990); McKenna et al. (1978); NTP 1985a, b |
| Aminoglycosides | Gram-negative antibiotics | Decreased GFR and increased serum creatinine and BUN; tubular necrosis | Klaassen (2001) |
| Ethylene glycol | Anti-freeze | Oxalate crystal formation; oxalate nephrosis; acute tubular necrosis; renal failure | Gordon and Hunter (1982); Mallya et al. (1986); Penumarthy and Oehme (1975); Siew et al. (1975) |
| Furosemide | Treatment of hypertension | Interstitial nephritis | Hardman and Limbird (2001) |
| Cyclosporine A | Transplant rejection; autoimmune diseases; skin disorders | Pyelonephritis; abnormal urine; hematuria; elevated BUN; polyuria; nocturia | PDR (2007) |
| Vancomycin[a] | Antibiotic | Interstitial nephritis (rare); increased BUN and serum creatinine | Farber and Moellering (1983); Hardman and Limbird (2001) |

[a]Total systemic and renal clearance of Vancomycin may be reduced in the elderly (FDA 2009)

## *2.2 Aging and the Liver*

The process of aging results in changes in hepatic structure and function. Physiologic changes occur during the aging process that can influence biotransformation reactions in the liver (Birnbaum 1991). A decline in liver size and hepatic blood flow, along with a reduction in metabolic capacity, has been demonstrated in older individuals (Anantharaju et al. 2002; Bianchi et al. 1988; Marchesini et al. 1988; Woodhouse and Wynne 1988; Wynne et al. 1989a; Zeeh and Platt 2002). Liver size decreases after age 50, paralleling a decrease in body mass until age 70 (Hazzard et al. 1999). A number of studies have shown a reduction in liver volume and blood flow on the order of 17–46% with advancing age (Marchesini et al. 1988; Woodhouse and Wynne 1988; Wynne et al. 1989c). Wynne (2002) found that individuals over 65 years of age had liver volumes 28% lower than individuals under 40. By age 80 years, hepatic mass has typically declined by about 40%, accompanied by a proportionate reduction in hepatic and splanchnic blood flow (Kampmann et al. 1975). The resultant reduction in the capacity for metabolism, biotransformation, and protein synthesis renders the liver less capable of responding to physiological stressors.

The aged body is less able to clear the body of xenobiotics (Zeeh and Platt 2002). It has been shown that the ability to metabolize some prescription medications is compromised in the elderly (Aramaki et al. 1998; Hazzard et al. 1999). The decrease in drug clearance appears to parallel the reduction in liver volume that accompanies aging (Bach et al. 1981; Swift et al. 1984).

A decrease in the rate of induction of hepatic microsomal activity (Hunt et al. 1992a, b; Woodhouse et al. 1984) has been reported to result in a decrease in benzodiazepine metabolism, narcotic clearance, aminopyrine demethylation, galactose elimination, and caffeine clearance in elderly patients. Drug metabolism mediated by the P450 system declines progressively after the fifth decade of life and undergoes another decrease in individuals over 70 years of age (Anantharaju et al. 2002). Other drugs known to undergo decreased hepatic metabolism in the elderly include ibuprofen, naproxen, meperidine, diltiazem, nifedipine, propranolol, quinidine, theophylline, verapamil, chlordiazepoxide, diazepam, imipramine, nortriptyline, trazodone, and levodopa (Beers et al. 2006). Changes in drug-metabolizing enzymes have also been reported in rats (Agrawal and Shapiro 2003; Jourdan et al. 2004; Spearman and Leibman 1984).

As a result of this age-associated reduction in hepatic capacity, Zeeh and Platt (2002) suggest that both renally excreted drugs and drugs metabolized and excreted by the liver should be applied at a starting dose 30–40% smaller than the average dose used in middle-aged adults.

There is no evidence to suggest that the ability of the body to metabolize and eliminate environmental chemicals would be different than for drugs, since the same enzyme systems are available for all xenobiotics. Because many environmental chemicals are metabolized by the P450 system (Hodgson and Rose 2007), it is also likely that the same structural and functional changes that may lead to increased susceptibility to adverse effects in drugs used in medical practice also result in increased susceptibility to insult from exposures to environmental chemicals. For example, Wynne et al. (1989b) provided evidence that the reduction in liver size in the elderly can affect the metabolism and clearance of acetanilide (*N*-phenylacetamide), a chemical with both manufacturing and medical applications.

Kinirons and O'Mahony (2004) reviewed the literature regarding changes in drug metabolism with aging. They concluded that age-associated reductions in the function of some, but not all, cytochrome P450 enzymes (CYPs) had been identified, but that there is considerable interindividual variability in drug metabolism with advancing age. Hepatic cytochromes 1A2, 2C9, 2C10, 2C18, and 2C19 have been shown to have reduced activity with age, and there is some evidence suggesting that a reduction in CYPs 2A, 2E1, 3A3, and 3A4 activity may also occur with aging (Kinirons and O'Mahony 2004). This reduction can have a negative impact not only on xenobiotic clearance, but also on the manifestation of toxicity from the extended presence of the parent compound. For example, the non-steroidal anti-inflammatory drug acetaminophen is oxidatively metabolized in the liver by the CYP enzymes 2E1, 1A2, 2A6, and 3A4 to the short-living *N*-acetyl-*p*-benzoquinone-imine, which is rapidly conjugated with glutathione and excreted renally (Farrell 2009). The possibility of age-related decreases in CYP 1A2 and possible decrease in CYPs 2E1 and 3A4 would certainly slow the breakdown of the parent compound, which could lead to hepatotoxicity in the elderly.

Age-related changes in hepatic metabolism can also result in clinically relevant, negative drug interactions between some pharmaceuticals (Dilger et al. 2000), as well as certain environmental chemicals. For example, ethanol can enhance the hepatotoxicity of acetaminophen, potentially to extremely toxic levels or even lethality (Lesser et al. 1985; Licht et al. 1980; Sinclair et al. 1998; Zimmerman and Maddrey 1995).

Other age-associated ultrastructural hepatocellular changes include a decrease in the number of mitochondria per hepatocyte, a decrease in endoplasmic reticulum, and an increase in the number of lysosomes, resulting in an increase in lipofuscin (Bianchi et al. 1988; Hazzard et al. 1999). These changes, particularly the reduction in endoplasmic reticulum, suggest a decreased ability of the aging liver to metabolize pharmacologic and environmental substances to which a senior citizen may be exposed.

In addition, as the lipid portion of the body increases, body fat may function as a reservoir for lipid-soluble pharmaceuticals and environmental chemicals, alike. Thus, the elderly may be more susceptible than are young adults to injury from lipid-soluble chemicals at the same exposure levels.

Several known hepatotoxicants are shown in Table 2.

## *2.3 Aging and the Cardiovascular System*

With aging, changes occur both within the heart and the vasculature. Despite the significant variability among individuals, the function of the cardiovascular system declines with age (Goldspink et al. 2003). The ability of the heart to pump blood around the body during maximal stress (called maximal cardiac power output) decreases in an age-dependent fashion (Cooke et al.1998; Williams et al. 2001).

**Table 2** Environmental/pharmaceutical agents causing hepatotoxicity

| Substance | Effect | Species | Reference |
|---|---|---|---|
| Acetaminophen | Centrilobular hepatic necrosis; hepatic encephalopathy; hepatic vascular disorders | Human/mouse | Farrell (2009); Jaeschke et al. (2002); Walker et al. (1983) |
| Arsenic | Cirrhosis; increased ALT, AST, ALP, bilirubin; hepatomegaly; vascular fibrosis | Human | Chakraborti and Saha, (1987); Franzblau and Lilis (1989); Guha Mazumder et al. (1988); Hernandez-Zavala et al. (1998); Kamijo et al. (1998); Levin-Scherz et al. (1987); Morris et al. (1974) |
| Carbon tetrachloride | Steatosis; increased serum bilirubin; increased serum ALT/SGOT/SGPT; centrilobular necrosis | Human/rat/mouse | Allis et al. (1990); Barnes and Jones (1967); Blair et al. (1991); Bruckner et al. (1986); Condie et al. (1986); David et al. (1981) |
| Cyclosporine | Elevated ALP, AST, ALT, bilirubin; cholestasis; cholangitis | Human/rat | Actis et al. (1995); De la Cruz Rodriguez et al. (2007); Kassianides et al. (1990); Lewis and Zimmerman (1999) |
| Dieldrin | Increased liver weight; elevated SER and P450 hemoprotein | Rat | Hutterer et al. (1968) |

**Table 2** (continued)

| Substance | Effect | Species | Reference |
|---|---|---|---|
| Estrogens | Cholestasis; jaundice; elevated ALP | Rat/mouse/monkey | Crocenzi et al. (2006); Germain et al. (2002); Kaplowitz et al. (1986); Leuenberger et al. (2009); Slikker et al. (1983); Yamamoto et al. (2006) |
| Ethanol | Fatty liver; hepatitis; cirrhosis; hepatocyte death | Human/rat/baboon | Fickert and Zatloukal (2000); Lieber and DeCarli 1976; Lieber et al. (1965); Worner and Lieber (1985) |
| Polychlorinated biphenyls (PCBs) | Fatty vacuolization/lipid accumulation; increased serum cholesterol and/or triglycerides; necrotic foci; increased liver weight; bile duct hyperplasia | Rat/mouse/monkey | Allen et al. (1974); Andrews (1989); Barsotti et al. (1976); Bruckner et al. (1973, 1977); Carter (1985); Carter and Koo (1984); Goldstein et al. (1974); Gray et al. (1993); Kato and Yoshida (1980); Koller (1977); Litterst et al. (1972); Price et al. (1988) |
| Tetracycline | Fatty infiltration; toxic hepatitis; acute and sub-acute necrosis; hepatic coma | Human | George and Crawford (1996); Heaton et al. (2007); Westphal et al. (1994) |
| Vitamin A | Vacuolization; non-cirrhotic portal hypertension; mild chronic hepatitis; cirrhosis | Human | Geubel et al. (1991); Kowalski et al. (1994); Minuk et al. (1987) |

The heart, like other organs, has a reserve capacity to ensure its proper function during stress (Cooke et al. 1998). Any decrease in the number of cardiomyocytes must inevitably impact that functional reserve of the heart. Olivetti et al. (1991) demonstrated that approximately one-third of cardiomyocytes are lost from the human heart between the ages 17 and 90, resulting in a decrease in pumping reserve and the ability to cope with exercise and life-threatening stresses (Tan and Littler 1990).

A thickening of the myocardium occurs with aging. This thickening is primarily due to an increase in the size of the individual myocytes. The amount of fibrous tissue within the myocardium increases with age, but this does not result in neovascularity (Morely and Reece 1989) and does not contribute appreciably to cardiac mass. Because mammalian cardiomyocytes have only limited regenerative capacity, cell death results in a net loss of viable contractile elements and a decrease in cardiac functional reserve during the course of aging (Goldspink et al. 2003). Some of the dead myocytes are replaced by fibrous connective tissue, which does not contribute to contractility or myocardial thickness.

Amyloidosis occurs in the hearts of some, but not all, elderly persons. In about half of those over 70, amyloid can be detected in the heart, with the incidence sharply increasing with advancing age. However, some elderly show no such presence of amyloid in the heart, even in centenarians (Beers and Berkow 2000).

Although systolic function is relatively preserved during aging, the duration of contraction is prolonged as the result of multiple factors (Lakatta 1993). There is a delay in diastolic relaxation, which affects ventricular filling throughout diastole. Diastolic dysfunction renders the aged individual to be substantially more dependent on atrioventricular synchrony and much more affected by tachycardia (Geokas et al. 1990).

Ventricular systolic stiffening and arterial stiffening occur with aging, even in the absence of cardiac hypertrophy. This stiffening results in a variation in cardiac filling, leading to disproportionately greater changes in systolic pressure in older individuals (Chen et al. 1998). In addition, stiffness of the myocardium resulting from myocyte hypertrophy and fibrosis, along with delayed ventricular relaxation, collectively leads to increased venous-filling pressures.

The inotropic, chronotropic, and vascular responsiveness of the sympathetic component of the autonomic nervous system to catecholamines also decrease in the elderly. However, excess catecholamine stimulation can be cardiotoxic (Benjamin et al. 1989). Ng et al. (2002) reported that single injections of a pharmacological dose of either noradrenaline (norepinephrine), adrenaline (epinephrine), or isoprenaline (a synthetic epinephrine derivative) to normal rat hearts resulted in both apoptosis and necrosis in cardiomyocytes. There were regional differences of cell death, with more than 10 times those effects being observed in the sub-endocardium than in the sub-epicardium of the left ventricle, suggesting different thresholds for injury for different regions of the heart (Ng et al. 2002).

During exercise, ventricular performance is compromised in the elderly (Stratton et al. 1994). These researchers reported an age-associated decline in heart rate, ejection fraction, and cardiac responses to supine exercise in healthy men. It has been proposed that the increased impedance of the central elastic vessels (arteries) with aging impairs ventricular performance, reduces the ventricular ejection fraction, and decelerates aortic flow, even in the absence of heart failure (Westerhof and O'Rourke 1995). This does not, however, mean that reasonable exercise cannot improve cardiovascular performance in most people as they age normally.

Le Page et al. (2009) found that regular exercise improved myocardial contractile function in senescent (24-month-old) male Wistar rats. And Stratton et al. (1994) reported that despite the significant cardiovascular changes that occur in the response to a single bout of exercise with aging, positive adaptations to chronic exercise training were not different with aging and included improvements in maximal work load and increases in ejection fraction and stroke volume at peak exercise.

The walls of the large arteries, such as the aorta, thicken, become dilated, and elongate with age. Systolic blood pressure increases by about 6.0–7.0 mm per decade due to an age-related progressive stiffening of the arteries (Rodriguez et al. 1994). Some changes emanate from a thickening of the intima as a result of cellular accumulation and to matrix deposition, along with fragmentation of the internal elastic membrane (Beers and Berkow 2000). Such is the case that occurs with fatty deposition (cholesterol and triglycerides) in the arterial walls leading to atherosclerosis. Other changes in arterial walls occur in the media (Nichols and O'Rourke 1998). During the aging process, elastic fibers undergo progressive disorientation, fragmentation, and degeneration, accompanied by subsequent collagen deposition, calcification, and/or cystic generation.

Although the walls of arteries are generally affected by the aging process, not all are affected to the same extent. In a study of 78 subjects (mean age 47 $\pm$ 6 years; range 23–71 years), 52 of whom had mild-to-moderate essential hypertension and 26 with no history of high blood pressure, Benetos et al. (1993) conducted non-invasive evaluations of blood pressure changes in the common carotid and femoral arteries. Results indicated that, although the carotid artery is very compliant in young patients, there was a strong decrease in the elastic properties of that artery with aging and increased blood pressure. The femoral artery was found to be much less compliant and less affected by aging and high blood pressure.

## 2.4 Aging and Hematopoiesis

Hematopoiesis is the process of synthesizing new blood cells of all types from pleuripotent stem cells in the red bone marrow. The production of red blood cells (erythrocytes) that carry oxygen necessary for cellular respiration in all tissues of the body is regulated by a hormone called erythropoietin, which is produced almost exclusively by the kidneys. Red blood cells have a limited life span, before they are broken down, primarily by the spleen. The turnover of red blood cells (RBCs) in a 70-kg adult is 200 billion per day (Ruscetti et al. 1998). This means that nearly two million die every second and an equal amount of RBCs are synthesized each second.

The aging process is accompanied by a reduction in hematopoiesis. Several causes for this phenomenon have been suggested, including a reduction in erythropoietin production, disease, increased apoptosis of hematopoietic stem cells, the aging process itself, and a reduction in lean body mass, which necessitates fewer hemoglobin-carrying RBCs to provide oxygen to the reduced muscle mass (Forbes and Halloran 1965; Lipschitz et al. 1984; Marley et al. 1999; Pearce and Bonnet 2009; Takafumi et al. 2000; Timaffy 1962).

Marley et al. (1999) provided evidence that the replicative capacity of myeloid progenitor cells declines with age and becomes more pronounced with advanced aging. This is accompanied by a decrease in the cellularity of bone marrow after the age of 80, along with a statistically significant increase in apoptosis of bone marrow cells in those over 80 years (Ogawa et al. 2000). The percentage of CD3-positive T cells and CD20-positive B cells in bone marrow has been reported to peak at age 60 and then decrease thereafter (Ogawa et al. 2000). There was also a decrease in macrophage density in adults and the elderly, compared with children. Ogawa et al. (2000) suggested that this decrease in the number of macrophages may have an influence on the reduction of hematopoietic cell proliferation and the induction of apoptosis in the bone marrow of elderly people and stressed the importance of the microenvironment in supporting and maintaining hematopoiesis in the bone marrow. This is supported by Wagner et al. (2008) who stated that aging is not only associated with functional alterations of hematopoietic stem cells, but also with an altered microenvironment that is required for hematopoietic differentiation.

Dietary factors also affect the ability of the body to produce RBCs to constantly replace dead or severely damaged red cells. In the process of synthesizing new cells, pleuripotent stem cells in the red marrow produce undifferentiated cells (called blast cells). Proerythroblasts, the type of blast cell that leads to the production of mature erythrocytes, then go through a number of developmental stages before being released into the blood as reticulocytes. Once in the blood, the reticulocytes become mature erythrocytes within a day or two. For this sequence of differential changes to proceed to conclusion, two water-soluble vitamins are essential. Vitamin B12 and folic acid are needed for the maturation of large, immature cells called megaloblasts, which synthesize thousands of hemoglobin molecules that will ultimately carry the oxygen to the body's tissues. Without B12 and folic acid, cell differentiation is unable to proceed, and megaloblasts, rather than reticulocytes, are released into the blood. The immature megaloblasts lack sufficient hemoglobin to adequately provide oxygen to the tissues and (megaloblastic) anemia results.

In nature, vitamin B12 (which is found only in foods of animal origin) has a co-factor as part of its structure. However, B12 cannot be absorbed through the intestinal wall with the co-factor attached. A type of stomach cell, called parietal cells, produces a substance called intrinsic factor, which cleaves the co-factor from the vitamin, thus enabling intestinal absorption into the blood to occur. With aging, there is a decrease in the number of parietal cells, resulting in compromised B12 absorption, thus leading to megaloblastic anemia.

Folic acid antagonist drugs such as methotrexate, which is used in cancer therapy and to treat rheumatoid arthritis, can also produce megaloblastic anemia as a result of their ability to inhibit DNA synthesis in the bone marrow (Klaassen 2001). Environmental substances (such as benzene) that cause hemolysis damage or inhibit stem cells in the bone marrow, or damage the kidneys (which release erythropoietin in response to low blood $O_2$), can thus have a greater negative impact on the elderly than young, healthy adults.

Further, exposure to environmental particulates, smoke, or other substances (by any route of exposure) may reduce hematocrit count. Xenobiotics known to be associated with anemia are shown in Table 3.

**Table 3** Environmental/pharmaceutical agents associated with anemia (by type)

| Chemicals causing megaloblastic anemia | Chemicals causing sideroblastic anemia | Chemicals causing aplastic anemia |
|---|---|---|
| Carbamazepine | Chelating agents | Benzene |
| Cholestyramine | Chloramphenicol | Bismuth |
| Colchicine | Cycloserine | Carbon tetrachloride |
| Ethanol | Ethanol | Chlordane |
| Omeprazole | Isoniazid | Dinitrophenol |
| *para*-Aminosalicylic acid | Lead | Gold |
| Phenobarbital | Pyrazinamide | Organic arsenicals |
| Phenytoin | Zinc | Mercury |
| Primidone | | Parathion |
| Sulfasalaizine | | Phenylbutazone |
| Triamterene | | Potassium perchlorate |
| Zidovudine | | |

## 2.5 Decreased Bone Density

A progressive decrease in bone density occurs during the aging process (Burger et al. 1998). This decrease is the combined result of decreased mineralization of the bone and a concomitant decrease in protein synthesis (Burger et al. 1998; Hannan et al. 1992; Jones et al. 1994; Yamaguchi and Ozaki 1990). Calcium is the primary mineral in bone tissue (Branca 1997). The absorption of dietary calcium declines with age (IOM 1997; Nordin et al. 2004). Collagen, which is the most abundant protein in the body, is necessary as a "scaffolding" around which mineralization can occur in bones (Bailey and Knott 1999). With advanced age, the production of bone collagen decreases, and molecular changes occur in some of the collagen that is produced (Bailey and Knott 1999).

Dietary practices among the elderly also contribute to the decrease in bone density. One of the best sources of calcium is milk, as well as other dairy products. Vitamin D is important in the absorption of calcium through the gut and is imperative for the absorption of calcium triphosphate (hydroxyapatite) into the bone. By law, all milk sold commercially in the United States is fortified with vitamin D. Milk, therefore, provides not only the calcium necessary for proper mineralization of bone tissue, but also the vitamin essential for the incorporation of that calcium into the bone. However, lactose intolerance is common among the elderly, and many chose to eliminate dairy products from their diet. Thus, a decrease in calcium in the diet, reduced absorption of dietary calcium, a decrease in the collagen production, defects in some of the newly synthesized collagen, and inadequate mineralization all contribute to an age-related increase in brittleness of bone tissue.

The decline in bone density is typically slow and steady. In men, it generally begins in mid-life (ages 60–65). In women, however, it typically begins in their 30s and increases significantly following menopause (Bonnick 1994). After age 35, women lose from 0.5 to 1% of their bone mass every year. Following menopause and the cessation of estrogen production by the ovaries, the rate of bone loss increases to 3–7% each year (in the absence of estrogen replacement therapy). By the end of the fifth year post-menopause, many women have lost from 15 to 35% of their peak bone density at 35 years of age (Bonnick 1994; Edelson and Kleerekoper 1995).

Xenobiotics can exacerbate the problems associated with the aging process. Caffeine, a widely consumed methyl xanthene, has been shown to increase the loss of calcium in the urine (Massey and Whiting 1993). Rapuri et al. (2001) found an association between caffeine intake levels > 300 mg/d with increased bone loss in a 3-year prospective epidemiological study; however, this bone loss was significant only for women having a particular genotype (*ttVDR*). Although younger individuals are able to compensate for these losses through increased calcium absorption, the elderly are less adaptable in this respect (Massey 1998).

The use of two of the most common environmental substances – alcohol and tobacco – increases the risk of developing osteoporosis in both men and women. Current cigarette smoking was found to be accompanied by a statistically significant increased rate of bone loss in both men and women (Burger et al. 1998). The reduction in bone density in the case of alcohol only occurs to a significant extent with heavy alcohol consumption (Sampson 2002); the effects of moderate or reduced drinking are less clear (Burger et al. 1998; Sampson 2002).

Accordingly, toxicants that are sequestered in the bone or that affect the process of bone mineralization or maintenance of calcium homeostasis may further compromise bone strength. For example, excessive exposure to lead or fluoride by the elderly would make compact bone more brittle and more susceptible to breakage during a fall or similar accident. Similarly, dermal, hepatic, and/or renal toxicants may decrease the production of vitamin D, which is essential to mineralization of bone tissue; and environmental xenobiotics that cause renal damage resulting in excessive calcium loss can also adversely affect bone density in both men and women. There is also evidence that some phthalates [benzyl butyl phthalate (BBP) and di-*n*-butyl phthalate (DBP)] induce apoptosis in the osteoblasts of rats and mice (Sabbieti et al. 2009), which certainly would impact the strength of bones.

## *2.6 Aging and the Nervous System*

The human brain contains approximately 100 billion neurons at birth, and that is the maximum number we will have at any point in our lives. There is compelling evidence that the brain shrinks with age (Abe et al. 2008), accompanied by a loss of synaptic contacts. Along with neuronal and volume loss, there is also an expansion of the extracellular space (Meier-Ruge et al. 1992). Gray matter, white matter, and cerebrospinal fluid compartments all decrease during the aging process (Abe et al. 2008; Allen et al. 2005; Good et al. 2001; Jernigan et al. 2001; Pfefferbaum et al. 1994; Raz et al. 1997; Salat et al. 2005; Walhovd et al. 2005). Gray matter accounts for most of the neuronal loss, while there is little concomitant loss of glial cells (Oskvig 1999). Eventually, 50% of the neurons of the cerebral and cerebellar cortices, locus ceruleus, thalamus, and basal ganglia have undergone apoptosis, and the remaining synaptic interconnections will have been markedly simplified (Feldman 1976). By the time a person reaches 80 years of age, the mass of the brain has decreased by about 20% from early adulthood and the volume of the cranial vault occupied by the brain decreases from 92 to 52%, with a compensatory increase in cerebrospinal fluid (Oskvig 1999).

To further investigate the loss of brain tissue that accompanies aging, Abe et al. (2008) examined the global and regional effects of aging on the brain in 73 normal female subjects (aged 22–70) using voxel-based volumetric analysis of brain by magnetic resonance imaging (MRI). They found that some areas of the brain showed a decrease in volume, while the volume in other areas was preserved. Bilateral, accelerated loss of volume was observed across widespread areas of the brain, particularly in anterior regions. The frontal, temporal, and parietal lobes, basal ganglia, and extranuclear white matter were all found to decrease in volume, whereas the bilateral cingulate gyri and subjacent white matter volumes were preserved (Abe et al. 2008). Globally, the negative association between age and brain volume was attributed almost exclusively to the shrinkage of gray matter. This finding was consistent with the results of previous studies of humans (Courchesne et al. 2000; Good et al. 2001; Pfefferbaum et al. 1994). White matter was observed to remain relatively stable until the age range of 60–70 years (Abe et al. 2008). Courchesne et al. (2000) looked at brain MRIs of subjects ranging in age from 19 months to 80 years, and they found that brain volume and intracranial space grew from early childhood through adolescence, and then decreased. The volume of gray matter continued to increase slowly to a plateau in the fourth decade. Courchesne et al. (2000) found that the gray matter-to-white matter ratio in healthy subjects declined after the fourth decade of life, and that subjects in the age range 71–80 had undergone brain volume decreases to levels similar to those that exist in young children.

Neuronal loss in the autonomic nervous system also progresses consistently with aging, resulting in a 15% loss of neurons by age 80 (Oskvig 1999). Baroreceptor, vasoconstrictor, and postural responsiveness are all impaired with aging (Oskvig 1999). Dizziness, unsteadiness, imbalance, and vertigo are common among the elderly (Nagaratnam et al. 2005). An inevitable decrease in short-term memory also occurs in the aging process.

With aging, the loss of neurons and a decreased capacity for sending nerve impulses to and from the brain results in the diminished processing of sensory information. Dorfman and Bosley (1979) compared nerve conduction velocities between 15 healthy young adults (mean age 31.6 years) and 15 "normal" elderly adults (mean age 74.1 years) and found that a reduction in conduction velocity began to be manifested around age 60. The efferent motor nerve conduction velocity was reported to subsequently decrease by 0.15 m/s/year. Initially, this might seem like an extreme decrease, but not after we compare this with normal motor nerve conduction velocities in healthy adults. Afferent nerve conduction mean velocities of 80.3, 67.5, and 54.7 m/s for median, ulnar, and tibial nerves, respectively, were reported for normal adult humans (Macefield et al. 1989). Efferent mean nerve conduction velocities of 56 m/s and approximately 55 m/s were reported for medial motor and peroneal motor nerves, respectively, in normal adult males (Buschbacher and Koch 1999). Thus, while a 0.15 m/s/year might not itself result in a significant compromise of neuromuscular function, such a decrement over a period of decades might result in a significant degradation of spinal reflexes and overall co-ordination. When this decrease in nerve conduction velocity in peripheral nerves is coupled with slower corticospinal transmission, the overall impact is slower initiation of voluntary motor activity and reflexes (Dorfman and Bosley 1979). With advanced aging, the nervous system becomes not only less efficient, but also less able to protect itself from exogenous influences. A primary interface between the brain and the systemic arterial circulation, which contains a myriad of nutrients, hormones, metabolic by-products, and proteins with a diverse array of functions, is the blood–brain barrier (BBB). The BBB is a morphological arrangement that allows the cerebral microvasculature to selectively protect the brain against the rapidly changing environment of the systemic circulation (Shah and Mooradian 1997). Because it serves homeostatic, nutritive, and communicative roles, any compromise in its integrity or function can result in dysfunction of the CNS (Banks et al. 2000).

Structurally, the BBB is composed of relatively impermeable capillary endothelial cells with tight junctions to prevent leakage into the brain. The basement membrane of these cells is continuous around the capillary walls, providing an additional barrier to entry for substances in the blood. In addition, astrocytes, a type of multifunctional glial cell, have extensions ("feet") that press against the capillaries, allowing only selective passage of some substances.

Normal neuronal function depends on a delicate chemical balance among neurons and their synaptic connections. The BBB provides for an intricate interplay between transport systems, receptors, and tight junction-specific antigens to ensure the delicate homeostasis of the brain environment. It plays a crucial role in the exchange and transport of nutrients and hormones to the brain and the export of metabolic end-products from the CNS (Shah and Mooradian 1997). Together with the blood–cerebrospinal fluid barrier (BCB), the BBB provides a safeguard for brain homeostasis. Any compromise in the structure or function of these barriers can contribute substantially to chemical-induced neurotoxicity (Zheng 2001).

Age-related changes in these barriers can be the result of either alteration in the carrier molecules or the physiochemical properties of the cerebral microvessels. With advancing age, the cerebral vasculature undergoes substantial changes, including significant alterations to the BBB (Shah and Mooradian 1997). These changes include an increase in permeability and the weakening of detoxification capability, efflux, and repair functions, resulting in a reduced ability to defend the brain

**Table 4** Chemicals that interfere with blood–brain barrier function[a]

| Chemicals | Mechanism of action | Reference |
|---|---|---|
| Bis-chloronitroso-urea (BCNU) | Disrupt barrier structure, followed by increased influx of chemical and | Abou-Donia et al. (1996); Brace et al. (1997); Chandra et al. (1999); Kochi et al. (2000); Nagahiro et al. (1991); Romero et al. (1996) |
| Cyclosporine A | increase in toxicity | |
| Dinitrobenzene | | |
| NMDA | | |
| Pyridostigmine bromide | | |
| Trinitrobenzene | | |
| Aluminum | Alter barrier functions, but do not directly damage BBB | Banks and Kastin (1989); Perl et al. (1980); Serot et al. (1997); Szumanska et al. (1993); Zheng et al. (1996, 1999) |
| Lead | | |
| Manganese | | |
| Mercury (inorganic) | | |
| Benzocyclobutene (BCB) | Biotransformation of xenobiotics at brain barriers as part of brain defense mechanism | Kalaria et al. (1987); Riachi et al. (1991); Strazielle and Ghersi-Egea (1999) |
| 1-Methyl-4-phenyl-1,2,3,6-tetrahydro-pyridine (MPTP) | | |

[a]*Source*: Modified from Zheng (2001)

from neurotoxicants (Zheng 2001). Thus, injury, disease, exposure to environmental agents, lifestyle, and nutritional habits may exacerbate the normal effects of the aging process on the brain. Table 4 provides a listing of some neuroactive chemicals and their effect on the BBB.

With advancing chronological age, degenerative changes and diseases of the sense organs increase sharply, and the senses of vision, hearing, taste, smell, and touch are all degraded (Schiffman 2007; Wolfson 2001). The thresholds of these senses, as well as those for pain and temperature, increase exponentially (Campbell et al. 1999; Oskvig 1999). As a result of both sensory and motor impairments with advancing age, head–eye co-ordination is also diminished and may contribute to dizziness and problems with maintaining balance experienced by the elderly (Proudlock et al. 2004). In a review of age-associated changes in gait, balance, and sensory function, Wolfson (2001) compared the age-associated changes with changes attributable to diseases. It was reported that all aspects of sensory function diminish with age, resulting in modest sensory changes in older patients. Since vision and hearing loss have a greater negative potential impact on safety, the remainder of the discussion of sensory degradation will be limited to those two senses. Vision is affected by changes in both ocular structure and the resultant changes in visual perception/performance. The decrements in visual perception can lead to more serious risks, such as in driving. The elderly typically undergo a progressive loss of peripheral vision, estimated to be up to one-third by 75 years of age and as much as 50% or so by age 90 (Schiffman 2007). Together with a decrease in the speed of light adaptation, this places the elderly at greater risk of getting into an accident while driving an automobile, especially at night.

Around age 65, visual acuity is significantly diminished on tests of visual acuity and contrast sensitivity (Schiffman 2007). Li et al. (2001) examined visual evoked potentials (VEPs) in 40 human subjects, ranging in age from 21 to 75 years, using vernier electrophysiologic testing. In those tests, it was found that the amplitude of vernier VEP waveforms was significantly reduced in subjects older than 60 years. In addition, the latent period from the vernier stimulus to the first negative wave peak was progressively prolonged with increasing age (Li et al. 2001).

A list of the ocular changes' effects of aging on vision is provided in Table 5.

**Table 5** Some age-related ocular changes that affect vision[a]

| Change | Effect on function/performance |
|---|---|
| Decrease in pupil size (senile miosis) | Brighter light needed for illumination of objects |
| Reduced accommodative power of lens | Presbyopia; need for reading glasses |
| Partial blockage of drainage network for aqueous humor | Increased intraocular eye pressure; potential damage to vision |
| Liquefaction of the gelatinous vitreous humor and separation from retina | Creates "floaters" (can be very distressing for some elderly) |
| Alterations in the structure of protein molecules in the cornea, lens, and vitreous humor | Generates glare |
| Decrease in lacrimation | Dry eye in some elderly individuals, resulting in reduced image clarity |

[a]*Source*: Schiffman (2007)

Hearing loss also accompanies the aging process; and while hearing loss is prevalent among the elderly, it is commonly undiagnosed (Bagai et al. 2006; Bogardus et al. 2003). Presbycusis (age-related hearing loss) is the most common cause of hearing loss in the United States and is typically gradual, bilateral, and characterized by high-frequency hearing loss (Bagai et al. 2006).

Two broad studies attest to the dramatic change in hearing with advancing age (Agrawal et al. 2008; Gopinath et al. 2009). Agrawal et al. (2008) reported the results of audiometric testing of 5,742 US adults ranging in age from 20 to 69 years as part of the National Health and Nutrition Examination Survey, 1999–2004. This study reported that in the 2003–2004 period, 16.1% of the adults (equivalent to 29 million Americans) had speech-frequency hearing loss. Approximately one-half of these individuals had unilateral hearing loss and half had bilateral hearing loss. High-frequency hearing loss was found in 31% of the study population (12% unilateral; 19% bilateral), equivalent to approximately 55 million Americans (Agrawal et al. 2008).

The second study reported the prevalence of age-related hearing loss in older adults as part of the Blue Mountains Study, a population-based study of sensory loss and other health outcomes (Gopinath et al. 2009). In the period 1997–2000, audiometric testing was performed on 2,956 individuals aged 50 years or older. All together, 33% of the study participants showed some level of hearing loss. Age-related hearing loss was more prevalent in men than in women for each decade younger than 80 years, and hearing loss was found to double for each decade of life in this study. Bilateral hearing loss was found in 28.7% of the men and 17% of the women aged 60–69 years. A history of working in noisy environment was associated with a 70–90% likelihood of any and moderate-to-severe hearing loss, respectively.

Helzer et al. (2005) measured hearing sensitivity in 2,052 subjects ranging in age from 73 to 84 years (mean age 77.5 years) and found hearing loss to be extremely common in this population. The prevalence of speech-frequency hearing loss was 59.9%, and the prevalence of high-frequency hearing loss was 76.9%. Hearing loss was most common among white men, followed by white women, black men, and black women. As with the two other studies described above, increased risk was associated with increasing age, white race, diabetes mellitus, vascular disease, smoking, and occupational noise exposure, among other factors (Helzer et al. 2005).

The causes of age-related hearing loss all depend on pathologic conditions along the sound transduction pathway. Contributions to loss of hearing may result from disruption of the auditory pathway anywhere from the pinna (outer ear) to the brain (Bagai et al. 2006). Although there is more than one way to classify hearing loss, a common way to break the causes down is to the classify the loss as either conductive, sensorineural, or mixed (Bagai et al. 2006; Yueh et al. 2003).

Conductive hearing loss is the result of changes in either the external or middle ear structures, thus preventing sound waves from reaching the fluids of the inner ear. Sensorineural hearing loss results from changes in the component structures of the inner ear. For example, changes in the cochlea or auditory nerve can prevent nerve impulses from being transmitted to the auditory cortex of the brain (Bagai et al. 2006). Changes in the cochlea can include the loss of supporting cochlear cells and the hair cells or loss of cochlear neurons (Schiffman et al. 2003). Degenerative neural changes have also been reported in central auditory pathways, including the brain stem and cerebral cortex (Schiffman et al. 2003). Moscicki et al. (1985) reported a hearing loss in 83% of 2,293 adults ranging in age from 57 to 89 years, with the majority of cases displaying a mild-to-moderate sensorineural hearing loss in the high-frequency range. Mixed hearing loss includes elements of both conductive and sensorineural losses (Bagai et al. 2006).

Causes of conductive hearing loss include cerumen impaction, perforated tympanic membrane, otitis media, otosclerosis, cholesteatoma, tumor, or disarticulation of the ossicular chain due to trauma. Causes of sensorineural hearing loss can be genetic or acquired, such as prolonged exposure to loud noises, exposure to ototoxic substances (e.g., aminoglycosides), inner ear infections, Meniere disease, and diabetes mellitus, among others.

With the loss of hearing in both speech and high frequencies, the ability to locate sound in a space becomes more difficult. In addition, the inability to hear or locate warning signals, such as fire alarms, emergency sirens, or honking horns, can put the elderly at increase risk of accidents.

With the virtual inevitability of at least some hearing loss during the advancing years, chemicals that are known to be capable of causing damage to the auditory components might then cause even greater damage than they would in healthy, young adults. For example, exposure to solvents, such as toluene, xylene, and trichloroethylene can be particularly damaging in the elderly. Similarly, some heavy metals, such as lead, can cause hearing loss through neurological damage. With the already compromised hearing in the elderly, excessive exposure to this metal may evoke more serious damage in an already hearing-compromised elderly adult. Theoretically, at least, any substance that can affect the nervous system can potentially illicit an effect on hearing.

Thus, the nervous system in the elderly adult is already significantly compromised, and any further insult by xenobiotics could exacerbate the effects of aging. Because of the age-induced compromise in CNS function, the effects of environmental chemicals might be manifested in the elderly at lower concentrations/intake levels than in healthy young adults.

If we look at the collective contribution of the effects of aging of all organ systems on neurologic function, the picture is potentially severe. The increased permeability of the blood–brain barrier with advancing years reduces the effectiveness of this barrier in protecting the brain from environmental neurotoxicants. When this functional decrement is combined with a reduced renal GFR and decreased metabolism (discussed previously), a longer blood half-life renders the elderly more likely to have neurotoxicants pass into the brain. This may mean that solvents and some metals may present a greater health risk to the elderly than to adolescent and adult population. Further, Muravchick (1996) reported that there is a predictable increased sensitivity of the elderly to inhaled and injected anesthetics, which is consistent and progressive, such that the dosing requirement for such drugs drops by nearly 30% by age 80. Given this, it is reasonable to suspect that inhaled solvents and other neuroactive xenobiotics may affect the aging brain at lower concentrations than the general population. Further, with increased use of neuroleptic pharmaceuticals among the elderly, the threshold for adverse neurologic effects to environmental substances may be decreased even more. Decreased neurologic performance, when combined with compromised cardiac function with advanced age, could increase the adverse potential of cholinesterase-inhibiting pesticides (ATSDR 1993, 2008) and neuroactive substances, such as mercury (ATSDR 1999), in exposed elderly individuals.

## 2.7 Aging and the Immune System

Aging is typically accompanied by immunosenescence, which is defined by Mocchegiani and Malavolta (2004) as the state of dysregulated immune function that contributes to the increased susceptibility to infections, cancer, and autoimmune disease observed in old organisms, including humans. During the process of aging, the cells and tissues of the immune system can be affected in many ways. But, while there are certainly changes in the cells themselves, not all cells in a population are necessarily affected (Horan and Ashcroft 1997). Although most age-related effects on the immune system are modest and do not compromise function in the basal state, the ability to withstand stressors is typically reduced, particularly when complicated by malnutrition and/or co-morbidity (Horan and Ashcroft 1997). Hawkley and Cacioppo (2004), after reviewing a number of studies regarding stress and immune function with aging, reported that research on stress in older adults provides evidence that stress contributes to effects that mimic, exacerbate, and possibly accelerate the effects of aging on immunity.

The decline of the immune system in the elderly includes both primary and secondary changes within the system. The primary change is the age-dependent intrinsic decline of immune responsiveness. The secondary changes result from underlying diseases and various environmental factors, including diet, drug intake, and physical activity. The consequences of these changes are increased susceptibility to infections, disease, the emergence of tumors, and an increase in autoimmune reactions (Wick and Grubeck-Loebenstein 1997).

Two types of immunity are recognized: innate (or natural) immunity and acquired (or adaptive) immunity. Acquired immunity combines highly specific antigen recognition and memory through genetic modification of lymphocytes and clonal expansion, respectively. Innate immunity is an immediate and fast response that can also guide aspects of the adaptive response (DeVeale et al. 2004). Aging is associated with a decline in adaptive immunity and increase in innate immune function (DeVeale et al. 2004). These age-related innate and adaptive immune changes could be decisive for healthy aging and survival (Delarosa et al. 2006). The vulnerability of the severely ill or injured elderly to develop systemic inflammatory response syndrome and multiple organ failure has also been linked to age-related changes in the immune system (Milberg et al. 1995; Moore et al. 1996).

An integral component of both the innate and acquired immune components is the presence of T lymphocytes, or T cells. The thymus gland is essential for the maturation of pre-T cells from the bone marrow. The thymus weighs about 15 g at birth and grows in size until puberty, when it achieves its maximum size of around 35 g (Boyd 1932; Kumar et al. 2005). Computed tomography (CT) measurements of glandular size have confirmed thymic growth with increasing age until puberty (Francis et al. 1985). It then undergoes progressive atrophy, or involution (Weksler and Hutteroth 1974), to a weight of 5–15 g in the elderly (Kumar et al. 2005). The age-related involution of the thymus is accompanied by the replacement of the gland's parenchyma by fatty tissue. Francis et al. (1985) found that in over one-half of the patients over 40 years of age examined using computed tomography, total fatty involution of the thymus gland had occurred; and fatty replacement of the gland was seen in nearly all patients by the late sixth decade. Thus, thymic degeneration is implicitly involved in the reduction in immune function with advancing age.

### 2.7.1 Innate (Natural) Immunity and Aging

Immune responses are triggered by the recognition of a limited diversity of microbiological products by cells of the innate component of the immune system (Pawelec et al. 1998). Naïve T-cells may exist in a quiescent state in the body for an extended period and may be subject to aging processes relevant to non-dividing cells (Linton et al. 1996).

Relevant changes in the innate component of the immune system include changes in macrophage function, polymorphonuclear lymphocytes (PMNs), natural killer (NK) cells, alterations in antigen-presenting cells (APCs), and response to cytokines (Mocchegiani and Malavolta. 2004; Pawelec et al. 1998).

#### Macrophages

Changes in macrophage function occur during aging (Pawelec et al. 1998). Some of these changes may be linked to NK cell dysfunction (Albright and Albright 1998), whereas others may be associated with neuroendocrine or other influences (De la Fuente et al. 1998). Macrophage function has been demonstrated to be diminished in elderly mice, when murine macrophages were subjected to a number of classical activating signals from a variety of agents, including IFN-gamma and lipopolysaccharide (LPS) (Ding et al. 1994; Yoon et al. 2004).

Macrophages have also been suggested to actively contribute to dysregulated immune function by their secretion of suppressive substances, prostaglandins (PGE) in particular (Pawelec et al. 1998). Dendritic cells, the main antigen-presenting cells, have been reported to be inhibited by PGE2 (Rieser et al. 1998). It has been further demonstrated that PGE2 directly inhibits T cells. In a study of healthy human subjects over 70 years of age, Goodwin and Messner (1979) found that mononuclear cells (including T lymphocytes) in the elderly may be more susceptible to such inhibition than T cells from young individuals. Further, T-cell function has been suggested as being more depressed than B-cell function (Roberts-Thompson et al. 1974).

#### PMNs

Neutrophils, a form of polymorphonuclear leukocyte, are on the first line of defense for invading microorganisms. Upon tissue injury, they migrate from the vasculature to injured tissue and phagocytize the invading pathogen or damaged tissue. Decreases in neutrophil function and superoxide production have also been observed in the elderly (Poligano et al. 1994). Studies of phagocytosis by neutrophils using opsonized bacteria or yeast have shown a significant reduction in phagocytic ability in the elderly (Butcher et al. 2001; Emmanuelli et al. 1986; Mege et al. 1988).

However, Butcher et al. (2001) demonstrated that the reduction in phagocytosis was due primarily to a reduction in the number of microbes ingested per neutrophil, rather than a reduction in the number of neutrophils showing phagocytic activity. In reviewing a large number of studies regarding the role of aging on neutrophil activity, Lord et al. (2001) reported that both Fe-mediated superoxide generation and phagocytosis are attenuated in the elderly, with Fe-effector response playing a major role in the age-related decline in neutrophil function.

### Cytokines

Cytokines include a diverse array of factors involved in immunoregulation (O'Shea 1997). Although all bind to receptors to produce a response, some have an excitatory function, while others inhibit. The interleukins are one group of cytokines. IL-10, which has been shown to down-regulate the function of antigen-presenting cells (APCs), is produced at higher levels in the elderly. Castle et al. (1997) reported significantly increased IL-10 production by peripheral blood mononuclear cells (PBMCs) in frail nursing home residents, when compared with controls. Similarly, Pawelec et al. (1997) found that T-cell clones aged in tissue culture as a longitudinal model of clonal immunosenescence produce much more IL-10 than cells from the same clones tested at a young age, and this increase in IL-10 was associated with a decreased capacity for IL-2 secretion. IL-2 is necessary for a number of immune functions, including the development of T-cell memory, production of immunoglobulins by B cells, and induction of the differentiation and proliferation of NK cells (Waldmann 2006; Waldmann and Tagaya 1999). Monocyte production of tumor necrosis factor alpha, a cytokine antagonist, has been shown to increase among the elderly (Born et al. 1995). Although these do not constitute all components of immune dysregulation that have been implicated to occur in the elderly, they do collectively provide evidence of compromised immune function with aging.

### B Cells

The aging process also results in changes in B-lymphocyte, or B-cell, development. Whereas B-cell production in the bone marrow continues throughout life, it is substantially decreased during the aging process in humans and other mammals (Ghia et al. 1996; Nunez et al. 1996; Zharhary 1988). But, unlike the involution of the thymus that results in a decline in T cells, there is no comparable change in the bone marrow. Despite the decrease in production in bone marrow, however, the total number of peripheral B cells remains constant, most probably due to an increased life span and subsequent accumulation of B cells with advancing age (Ghia et al. 2000; Kline et al. 1999).

A decrease in the diversity of the peripheral B-cell repertoire has been observed in old mice (LeMaoult et al. 1997; Zharhary 1988), but serum immunoglobulin concentrations do not dramatically decline with age (Ghia et al. 2000). Although immune responses to T-cell independent antigens are comparable in old and young mice, antibody responses to T-cell dependent antigens have been shown to decrease dramatically (LeMaoult et al. 1997). The same decrease in the diversity of peripheral B cells also occurs in humans (Ghia et al. 2000).

Bone marrow production of different B-cell lineages declines with age in humans. For example, the number of total $CD19^+$ and $CD10^+$ B-lineage cells markedly decreases with age (Ghia et al. 1996; Nunez et al. 1996; Rego et al. 1998). The B1 subset of lymphocytes expressing CD5 receptors, which are abundant in neonates, decrease in adults. With aging there is also a shift from foreign to self-antigen specificities, paralleling a shift of producing cells from $CD5^-$ to $CD5^+$ (LaMaoult et al. 1997). Thus, aging brings an increased risk of autoimmune diseases. Further, specific antibody responses to vaccines have been found to decrease in older humans (Ben-Yehuda and Weksler 1992; Beyer et al. 1989). Colonna-Romano et al. (2006) reported lower serum IgD levels and higher $CD19^+$ $CD27^+$ memory cells in old individuals, when compared to younger subjects, suggesting that the B-cell repertoire available to respond to antigen challenge shrinks in the elderly, along with the number of naïve $IgD^+$ B-cells.

### 2.7.2 Acquired (Adaptive) Immunity and Aging

The aging immune system is less capable of coping with infectious disease than the youthful immune system (Pawelec et al. 1998). The elderly are more susceptible to all types of infection and malignancies resulting from the decrease in immune competence that occurs as a result of age-related changes. Although there is a progressive decline in both cell-mediated and antibody-mediated immune responses with age, the T lymphocytes are more severely affected than B cells (Grubeck-Loebenstein 1997; Miller 1996). Much of the decrease in T-cell population can be attributed to the aforementioned involution of the thymus, which is almost complete by the age of 60, making the aging individual dependent on the T-cell pool generated earlier in life (Grubeck-Loebenstein 1997).

This age-related decline in T-lymphocyte population includes a decrease in the number of helper T-cells, with a significant loss occurring between the ages 65 and 70. Horan and Ashcroft (1997) reported an impaired production of the IL-2 by helper T-cells, as well as a decreased responsiveness to that cytokine with advanced aging.

Further, B-cell production is impaired, resulting in a decrease in antibody production and a shortened immunological memory (Grubeck-Loebenstein 1997). This results in a decreased response to vaccines (Beyer et al. 1989; Globerson 1995; Gross et al. 1995; Naylor et al. 2005). In addition, there is an increase in the production of antibodies against self-proteins. This increase may be significant in some individuals, contributing to the aforementioned increased incidence of autoimmune diseases among the elderly.

To initiate an adaptive immune response, T cells must be activated by functional antigen-presenting cells or APCs (Pawelec et al. 1998). Thus, any alterations to APCs may have a significant impact on the adaptive immune response. In studies of mice, it was found that antigen-presenting macrophages from old mice stimulated lower levels of T-cell proliferation than macrophages from young mice (Kirschmann and Murasko 1992; Vetricka et al. 1985).

Dendritic cells play a pivotal role in co-ordinating immune responses to infection (Jones et al. 2006). They are antigen-presenting cells, central to the induction of immune responses. Antigen contact triggers their maturation, and they migrate to draining lymph nodes where they potently activate the proliferation of naive T-cells (Jones et al. 2006). Miller et al. (1994) found that dendritic cells in germinal centers of aged mice may lack expression of important co-stimulatory ligands such as CD86, a major co-stimulatory ligand, encouraging a state of immune unresponsiveness in antigen-specific T-cells. Steger et al. (1997) reported that dendritic cells from the elderly may fail to cross tissue barriers properly and have an impaired capacity to trigger IFN and IL-10 production by influenza-specific T-cells in vitro. In mice, defects in the transportation of antigens by dendritic cells to germinal centers of lymph nodes have also been reported (Holmes et al. 1984; Szakal et al. 1988).

Goodwin et al. (2006) performed a quantitative review of 31 vaccine antibody response studies conducted from 1986 to 2002 and compared antibody responses to influenza vaccine in elderly and younger adults. They concluded that the antibody response in the elderly was considerably lower in adults of age 65 and older and projected a clinical vaccine efficacy of about 17–53% in the elderly for the H1, B, and H3 antigens. Other recent studies have reported an age-related decline in the function in a variety of T-cell sub-sets (Deng et al. 2004; Kang et al. 2004; Murasko et al. 2002; Saurwein-Teissl et al. 2002). Murasko et al. (2002) reported that 35–50% of elderly subjects (ages 67–95 years) administered influenza vaccine for 4 years demonstrated neither an antibody nor cell-mediated response to the vaccine each year, attesting to the compromised state of the immune system in that population.

In consideration of the various ways in which the immune system is degraded in elderly individuals, elevated concern should be given by public health officials to exposure to immunotoxic xenobiotics, which can contribute to an already substantially compromised immune function. Thus, exposure to chemicals such as benzene and 2,3,7,8-tetrachlorodibenzo-*p*-dioxin (TCDD), the prototype of xenobiotic immunotoxicants (Klaassen 2001), should be considered of particular concern to the elderly. TCDD is known to be particularly toxic to the immune system of mammals. A number of studies have shown that this dioxin causes a reduction in thymic weight (i.e., causes thymic atrophy) in a variety of mammals, including monkeys (McConnel et al. 1978), rats (De Heer et al. 1994; De Wall et al. 1992; Hanberg et al. 1989; Murray et al. 1979; Van Birgelen et al. 1995; Viluksela et al. 1994; Vos et al. 1973), mice (Diliberto et al. 1995; Silkworth et al. 1989), and guinea pigs (DeCaprio et al. 1986; Hanberg et al. 1989; Umbreit et al. 1985; Vos et al. 1973). Significantly reduced polymorphonuclear leukocyte activity (Ackermann et al. 1989), antibody response (Holsapple et al. 1986), cytotoxic T-cell activity (De Krey and Kerkvliet 1995), and serum complement activity (Lin and White 1993; White et al. 1986) have been reported in mice exposed to TCDD. Decreased cell-mediated immunity (Fan et al. 1996; Vos et al. 1973), inhibition of thymocyte maturation (Blaylock et al. 1992), and suppressed humoral activity (Vecchi et al. 1983) have also been reported in TCDD-exposed mice. Compromised cell-mediated immunity (Fan et al. 1996) has also been reported in rats, and decreased lymphocyte activity has been seen in guinea pigs (Vos et al. 1973). Among primates, lymph node atrophy has been reported in monkeys (Allen et al. 1977), and overall immune suppression has been reported in humans exposed to TCDD (Tonn et al. 1996).

Benzene is another extensively studied chemical that has significant effects on the immune system. Both leucopenia and lymphopenia have been reported in humans (Cody et al. 1993; Kipen et al. 1989; Xia et al. 1995) and rats (Dow 1992; Li et al. 1986; Snyder et al. 1978, 1984; Ward et al. 1985; Wolf et al. 1956). These effects have also been reported in studies of mice, along with decreased thymic weight, decreased bone marrow cellularity, and decreased granulopoietic stem cells (Aoyama 1986; Chertkov et al. 1992; Cronkite 1986; Cronkite et al. 1985, 1989; Gill et al. 1980; Robinson et al. 1997; Snyder et al. 1978, 1980, 1988; Toft et al. 1982; Ward et al. 1985; Wells and Nerland 1991), demonstrating a remarkable similarity of effect across mammalian species.

Although TCDD and benzene have been extensively studied for their immune effects, exposure to a variety of metals, pesticides, and chlorinated organic hydrocarbons has also been reported to negatively impact the immune system. Table 6 contains a listing of some of these chemicals. This listing is far from complete, but is intended to demonstrate the variety of environmental xenobiotics that can negatively affect the immune system.

**Table 6** Environmental xenobiotics known to be immunotoxicants

| Chemical | Immunotoxic effect(s) | Species | Reference |
|---|---|---|---|
| Cadmium | Decreased humoral immune response; reduction in spleen lymphocyte viability and number; decreased suppressor cell activity; induction of anti-nuclear autoantibodies | Mouse | Blakley (1985); Graham et al. (1978); Krzystyniak et al. (1985); Malave and de Ruffino (1984); Ohsawa et al. (1988) |
| Dieldrin | Suppressed humoral immune response to both T-cell dependent and independent antigens; depressed macrophage function | Mouse | Fournier et al. (1988); Krzystyniak et al. (1985); Loose et al. (1981) |
| Dimethoate | Decreases in: number of rosette- and plaque-forming cells; response to T-cell and B-cell mitogens; thymus and spleen weight; antibody production; humoral and cellular immune response | Mouse/rat | Aly and El-Gendy (2000); Institoris et al. (1999); Tiefenbach and Lange (1980) |

**Table 6** (continued)

| Chemical | Immunotoxic effect(s) | Species | Reference |
|---|---|---|---|
| Mercury (inorganic) | Anti-nuclear antibody production; suppressed lymphoproliferative response to T-cell mitogens | Mouse | Dieter et al. (1983); Warfvinge et al. (1995) |
| Polychlorinated biphenyls (PCBs) | Reduced IgM and IgG antibody responses to sheep red blood cells (SRBCs); decreased anti-SRBC hemolysin titers | Monkey (Rhesus; Cynomolgus) | Thomas and Hinsdill (1978); Truelove et al. (1982); Tryphonas et al. (1989) |
| | Increased susceptibility to Moloney leukemia virus; thymic atrophy; decreased NK cells; increased sensitivity to bacterial endotoxin; decreased gamma-globulin-containing cells in lymph nodes; decreased antibodies to tetanus toxin and skin reactivity to tuberculin | Mouse, rat, guinea pig, or rabbit | Koller (1977); Smialowicz et al. (1989); Street and Sharma (1975); Thomas and Hinsdill (1978); Vos and de Roij (1972); Vos and Van Dreil-Grootenhuis (1972) |

## 2.8 Aging and the Endocrine System

As with other organ systems, the endocrine system undergoes a progressive loss of reserve capacity with aging. The reserves are gradually degraded and/or depleted by aging itself or intercurrent pathological states (Perry 1999). Decrements in hormone synthesis, metabolism, and response to hormones occur (Chahal and Drake 2007), but the effects of these changes are not normally apparent under baseline conditions (Hazzard et al. 1999). Blood levels of some hormones decrease, while some increase or remain unchanged (Chahal and Drake 2007). Many hormone level changes originate in the hypothalamus and/or pituitary gland.

The hypothalamic–pituitary–thyroid axis undergoes physiological alterations associated with the aging process. Many of those changes seem to be subtle and suggestive of a decreased hypothalamic stimulation of thyroid function (Leitolf et al. 2002). This is consistent with the substantial evidence that the hypothalamus becomes increasingly dysfunctional with aging.

The thyroid gland itself undergoes several anatomic changes with aging. The size of the gland itself decreases, as do the size of the follicles and the content of the colloid. With aging, there is a reduction in tetraiodothyronine ($T_4$) secretion, which is believed to be secondary to a reduction in $T_4$ clearance (Oddie et al. 1966). However, serum and free triiodothyronine ($T_3$) does decrease with age, due to a decreased peripheral conversion of $T_4$ to $T_3$ (Mariotti et al. 1995). The overall reduction in iodated thyronines is most certainly a contributor to the reduced metabolic rate experienced in aging. This may have significant impact in advancing age, and it has been suggested that even mild variations in thyroid function can have significant consequences on cognitive function in the elderly (Begin et al. 2008). A listing of chemicals known to affect the uptake of iodine into follicular cells is contained in Table 7.

**Table 7** Chemicals affecting the active uptake of iodide into thyroid follicular cells

| Chemical | Source/use | Mechanism | Reference |
|---|---|---|---|
| Dysidenin and its metabolite isodysidenin | Source: Sponge Dysidea herbacea | Unknown | Van Sande et al. (1990) |
| *N*-Substituted anthranilic acid derivatives | | Inhibition of chloride channels (reversible) | Fanelli et al. (1995) |
| Propranolol | Anti-hypertensive | Membrane stabilizing activity | Murakami et al. (2004) |
| Perchlorate | Manufacture of rocket propellant; other industrial uses | Competitive inhibition of iodide follicular transport | Clewell et al. (2004); NAS (2005) |
| 1-Methyl-2-mercapto-imidazole (methimazole; Tapazole) | Management/treatment of hyperthyroidism | Inhibition of active transport of inorganic iodide | Freinkel and Ingbar (1955) |
| Thiocyanate | Metabolite of cyanide; also in goitrogens and cigarette smoke | Inhibition of active transport of inorganic iodide | Erdogan (2003); Ghorbel et al. (2008); Kreutler et al. (1978); Steinmaus et al. (2007) |

With aging, there is an increase in the production of parathyroid hormone (PTH), which may contribute to the loss of bone density previously discussed (Endress et al. 1987).

During the aging process, there are changes in the hypothalamic–pituitary–adrenal axis, which have substantial physiological impact. Age has been linked to higher basal cortisol levels (Van et al. 1996). Van and colleagues (1996) reported a 20–50% increase in 24-h mean cortisol levels in individuals between the ages 20 and 80. Increased cortisol levels have been shown to be associated with decreased bone mineralization in men (Dennison et al. 1999) and risk of bone fractures in both men and women (Greendale et al. 1999). Increased cortisol levels have also been implicated in decline of memory (Wolf et al. 2005), sleep disorders, hippocampal atrophy, cognitive impairment, and sleep disorders among the elderly (Seeman et al. 1997; Van et al. 1996).

During aging, there is also a change in the performance of the renin–angiotensin–aldosterone system (Bauer 1993; Belmin et al. 1994). Specialized cells in the distal tubule of the renal nephron are sensitive to low blood pressure/volume and release of the hormone renin. Renin then acts on a plasma protein called angiotensinogen, converting it to angiotensin I. Angiotensin I is subsequently converted by angiotensin-converting enzyme (ACE) to angiotensin II in the lungs. Angiotensin II has a vasopressor effect and also signals the adrenal cortex to release aldosterone, which causes the reabsorption of sodium and the elimination of potassium.

It is well known that the plasma levels of renin and aldosterone are reduced with advanced age (Bauer 1993; Belmin et al. 1994). The basal renin level is reduced by 30–50% in the elderly, is accompanied by a comparable reduction in aldosterone, and is believed to be secondary to the renin reduction (Beers and Berkow 2000). There is also a decreased responsiveness of the adrenals to angiotensin II, contributing to the decrease in plasma aldosterone (Belmin et al. 1994). These reductions predispose the elderly to an increased risk of hyperkalemia in various clinical settings (Beers and Berkow 2000).

With aging, there is a reduction in the secretion and serum concentration of growth hormone (GH) and insulin-like growth factor-I (IGF-I) (Corpas et al. 1993). GH is involved in a number or physiological processes and has both anabolic and lipolytic actions. IGF-I, secreted primarily by the liver, is involved in the mediation of GH activity. The production of GH and its concentration decreases by more than 50% in healthy older adults (Veldhuis et al. 2005). Among the elderly, this decline in GH secretion is known to cause a result in a reduction in protein synthesis, lean body mass, bone mass, and immune function. The decrease in IGF-I levels are believed to result from the reduction in GH secretion, rather than a loss of hepatic response to the presence of GH (Corpas et al. 1993). There is evidence suggesting that the reduction in GH is, at least in part, due to the age-dependent decrease in the production of growth hormone-releasing hormone (GHRH) by the hypothalamus (Russel-Aulet et al. 1999).

Some of the more pronounced changes that occur during the aging process are related to changes in sex hormone production. Following both menopause and andropause, the endocrine system no longer functions as optimally as it did in younger adulthood. Changes in gonadosteroid production in older men and women result in a variety of effects, including a decreased mineralization of the bone, loss of libido, and resistance to insulin produced in the pancreas (Abate et al. 2002; Gray et al. 1991; Riggs and Melton 1986; Nordin et al. 2004). Although the onset of symptoms of decreased sex hormones begins around the age of 40 in women, it does not begin at any specific point in men and is different from the sharp reduction of estrogen production in females at the menopause, and may vary between modest and severe (Ishimaru et al. 1977; Johnson 1998; Vermeulen 2001). The various effects of the changes in sex hormone release in women and their causes and general time of onset are given in Table 8.

**Table 8** Peri- and post-menopausal changes in women

| Change | Effect | Reference |
|---|---|---|
| ***Pre-peri-menopausal transition:*** Decrease in serum estradiol; increase in FSH (typically by age of 40) | Decrease in frequency of ovulation (by age of 40); onset of bone loss | Johnson (1998); Sherman et al. (1976); Riggs and Melton (1986) |
| ***Peri-menopausal–menopause transition:*** Ovarian follicular activity ceases; estrogen falls to post-menopausal values; increase in FSH and LH above pre-menopausal values; periodic surges of LH; hypothalamic dysfunction; increased serotonin release (mid-fifth to mid-sixth decade) | Decrease in bone density 5–15%, primarily in trabecular bone; increase in LDL and total cholesterol; decrease in HDL cholesterol; increased cardiovascular risk; hot flashes; narrowing of thermoregulatory system; cognitive disturbances | ACOG (2004); Johnson (1998); Nordin et al. (2004); Riggs and Melton (1986); Speroff et al. (1999) |
| ***Post-menopausal period:*** Follicular activity has ceased; estrogen levels remain low; small amounts of estrone synthesized from androstenedione in the adrenal cortex and interstitial ovarian cells are transformed into estradiol; increased sensitivity to parathyroid hormone (PTH) | Atrophy of vaginal mucosa; vaginal bleeding; loss in libido; continued decrease in bone density; decreased temperature tolerance range; increased risk of coronary artery disease, myocardial infarction, and stroke | Johnson (1998); Longcope et al. (1980); Nordin et al. (2004); Speroff et al. (1999) |

During andropause, the male age-related version of menopause, there is a gradual, progressive decline in testosterone levels with advancing age (Gray et al. 1991; Harman et al. 2001; Morely et al. 1997). This drop results primarily from a decreased rate of testosterone production in older men (Ishimaru et al. 1977) and has been associated with abnormalities at all levels of the hypothalamic–pituitary–testicular axis (Mulligen et al. 1997). In addition, the response of testosterone to LH and human chorionic gonadotrophin decreases with age (Harman and Tsitouras 1980). The ultimate effects of these changes have been associated with decreased bone density, decreased muscle mass, increased body fat, increased cardiovascular risk, insulin resistance, anemia, poor libido, erectile dysfunction, and depression (Abate et al. 2002; Gray et al. 1991; Hak et al. 2002).

In addition, the blood level of dehydroepiandrosterone (DHEA) and its more common sulfate form (DHEAS) peak around age 20 and then begin to rapidly decrease around age 25. By the age of 80, DHEA levels have dropped to only

10–20% of those of younger adults (Orentreich et al. 1984; Vermeulen 1995). Although the physiological consequences of this decrease are not fully understood, associations between the decline in DHEAS and cardiovascular disease, breast cancer, reduced bone density, depressed mood, and type 2 diabetes mellitus have been reported. However, a causal relationship for these associations has not yet been established (Gurnell and Chatterjee 2001).

With aging, the epidermal cells of the skin become less capable of contributing to in vivo vitamin D synthesis. The level of epidermal 7-dehydrocholesterol, the starting point for vitamin D synthesis, has been reported to decrease in a liner fashion by about 75%, between early and late adulthood (Holick et al. 1989). In a study population of 1,606 community-dwelling men, 65 years of age or older, Orwoll et al. (2009) conducted serum assays for 25-hydroxyvitamin D (25-hydroxycalcitriol), an intermediate form of vitamin D synthesized (hydroxylated) in the liver from UV-irradiated cholesterol molecules from the skin. Since this form is more easily measured than 1,25-dihydroxycalcitriol, the ultimate form of vitamin D produced in vivo by a final hydroxylation of 25-hydroxycalcitriol in the kidneys, the monohydroxylated form is used in serum analyses of vitamin D (IOM 1997). Orwoll and his colleagues found a serum vitamin D deficiency (defined as <20 ng/mL) in 26% of the men tested and an insufficiency (<30 ng/mL) in 72% of the men. This can be compared to a maximal vitamin D serum concentration of approximately 60 ng/mL, resulting from UV exposure (Binkley et al. 2007). The deficiency reported by Orwoll et al. (2009) was particularly common among men during the winter and spring months, especially in northern communities. In Caucasian men in winter or spring who were >80 years of age and who did not engage in lawn or garden work and had a body mass index greater than 25 kg/m$^2$ and vitamin D intake below 400 IU/d, the prevalence of vitamin D deficiency was 86% (Orwoll et al. 2009). In a study population of 1,234 men and women aged 65 or older in the Netherlands, approximately 48% had serum vitamin D levels of 20 ng/mL or less, and approximately 82% had serum vitamin D levels below 30 ng/mL (Wicherts et al. 2007).

Agents that affect the quality of the epidermis and subsequently the production of vitamin D can place the elderly at increased risk of bone fractures.

## 2.9 Aging and the Integumentary System

Skin aging is a continuous process that affects both skin function and appearance (Li et al. 2006). Age-related changes in the skin are the result of both intrinsic and extrinsic factors (McCullough and Kelly 2006). A hallmark of aging is the progressive accumulation of molecular damage in nucleic acids, proteins, lipids, and other macromolecules (Tavernarakis and Driscoll 2002).

Changes in protein synthesis occur during the aging process (Syntichaki and Tavernarakis 2006). These changes have a far-reaching impact on both structure and function, and not all proteins are uniformly affected (Park and Prolla 2005).

Nonetheless, the reduction in protein turnover/replacement leads to an inexorable deterioration of essential cellular functions (Syntichaki and Tavernarakis 2006). Goukassian et al. (2000) reported data which suggest that an age-associated decrease in DNA-damage repair results from decreased availability of many proteins that participate in the repair process.

In addition to the decrease in protein synthesis, the aging process is also associated with damage to existing proteins (Balin and Allen 1989). Specific damage includes substitution of D-amino acids for L-amino acids within proteins. Such amino acid racemization is known to alter protein function. In addition, reducing sugar aldehydes consolidate with amino acid groups, resulting in loss of function (Balin and Allen 1989).

A major change in the loss of skin tissue with age is a deterioration of the dermal matrix. This change is related to the functional degradation of matrix rather than to a decrease of matrix synthesis (Ravelojaona et al. 2008).

Fibroblasts in the matrix produce collagen, elastin, and adipocytes and all are essential to the health of the skin. Collagen is a protein that gives strength to the skin, and elastic fibers, comprised of elastin, give resilience to the skin. Reduced synthesis of collagen types I and III is characteristic of chronically aged skin (Varani et al. 2006). In a study of skin aging in healthy adult volunteers ranging in age from 6 to 84 years, El-Domyati et al. (2002) found that collagen fiber architecture in facial skin became disorganized and underwent decrease in staining intensity after the fourth decade of life. Varani et al. (2006) attributed the reduction in collagen synthesis seen in chronically aged skin to cellular fibroblast aging (resulting in a reduced capacity for collagen synthesis) and a lower level of mechanical stimulation resulting from a decreased number of intact collagen fibers.

Elastin fibers in facial skin have been reported to be morphologically abnormal and appeared to occupy areas of lost collagen (El-Domyati et al. 2002). An age-dependent increase in matrix-degrading enzymes has also been demonstrated in human skin fibroblasts (Labat-Robert et al. 1992).

A reduction in the number of fibroblasts in the dermal layer occurs, accompanied by a decrease in blood vessels, mast cells, and neural elements that collectively decrease tensile strength, elasticity, and the ability to thermoregulate. A loss of fat from the sub-cutaneous layer (Rabe et al. 2006) further contributes to a compromised ability to maintain body temperature. Dermal fibroblasts, essential for dermal health and wound healing, lose both proliferative capacity and the ability to migrate during the process of wound healing (Ashcroft et al. 1995). Aged fibroblasts have also been shown to produce less matrix, resulting in a decrease in dermal tissue (Colige et al. 1990).

Aged cells have been shown to have lowered levels of epidermal growth factor receptors (EGFRs) on dermal fibroblasts (Reenstra et al. 1996; Shiraha et al. 2000). A decrease and delay in the number of occupied receptors that are transported intracellularly have been reported (Reenstra et al. 1996). Receptors for fibroblast growth factor and platelet-derived growth factor have also been shown to be reduced during the aging process (Aoyagi et al. 1995; Ashcroft et al. 1997; Garfinkel et al. 1996a, 1996b).

Thus, with the loss of sub-cutaneous fat, reduction in fibroblast production, and misregulation of normal collagen and elastin production comes the wrinkling and sagging of the skin seen in the elderly (Lange and Schnohr 2007; Rabe et al. 2006).

However, other age-related changes in the skin present a greater danger to health. With aging, the skin typically undergoes a degenerative process that contributes to a decline in its effectiveness to function as a barrier to the entry of living and non-living pathogens.

The time between the production of a new keratinocyte from the basal stem cells and the sloughing of the dead epidermal cell is approximately 28 days in young adults, but increases to 40–60 days in the elderly (Grove and Kligman 1983). This results in rougher, scalier, and more transparent skin. Yet, as we age, there is a reduction in the thickness of the dermis and epidermis, a flattening of the epidermal–dermal junction, and a decrease in the amount of sub-cutaneous fat (Hazzard et al. 1999; Montagna and Carlisle 1979; Moragas et al. 1993; Yaar and Gilchrest 2001). In males, the flattening of the epidermal–dermal junction is relatively constant throughout adulthood, whereas the decline in junction flatness in females occurs sharply between the ages 40 and 60 years, presumably related to the menopause (Moragas et al. 1993).

There is a decrease in the number and function of melanocytes with aging (Swift et al. 2001). This decrease amounts to approximately 10–20% of the melanocyte population each decade (Gilchrest et al. 1979). The reduction in melanocyte activity leaves the aging individual more susceptible to UV radiation-induced damage to DNA (McCullough and Kelly 2006), and it would seem likely that chemical agents causing photosensitivity would amplify this risk. The polycyclic aromatic hydrocarbons (PAHs) such as anthracene, fluoranthene, acridine, and phenanthrene are examples of such environmental chemicals (Klaassen 2001).

In addition, changes in the vasculature of the skin occur with aging, affecting the ability of the skin cells to be supplied with oxygen and other nutrients. The number of capillary loops in the dermal papillae decreases with age (Li et al. 2006). Kelly et al. (1995) reported a reduction in dermal papillary loops of 40% in the forehead and 37% in the forearm. This age-dependent reduction in papillary loop microvessels is accompanied by a decreased thickness of microvessel basement membranes and a decrease in the number of perivascular cells (Braverman and Fonferko 1982). These structural alterations have been implicated as obvious causes of decreased perfusion and increased capillary fragility associated with cutaneous aging (Chang et al. 2002). Since horizontal plexuses situated in the middle and lower layer of the dermis are related to skin temperature (Braverman 2000), these changes in the microvasculature of the skin can have significant effect on skin temperature.

The epidermis has been shown to decrease in thickness with advancing age. This decrease has been reported to be slightly faster in men (7.2% of the original value per decade) than in women (5.7% per decade), while the total dermal thickness decreases at a rate of 6% per decade in both men and women (Branchet et al. 1990). Thinning of the epidermis and the loss of up to 20% of the dermal thickness contributes to the appearance of paper-thin skin in the elderly. The remainder of the dermis is largely avascular and acellular (Hazzard et al. 1999). The thinning of these layers of the integument is largely related to a decline in the number of stem cells that renew these tissues, with a resultant decrease in wound healing capacity and cytokine, growth factor, and vitamin D production (Hazzard et al. 1999; Montagna and Carlisle 1979).

There is also a decrease in the number of epidermal Langerhans cells, which are instrumental in activating T lymphocytes and other lympho-immune cells in the integument (Gilchrest et al. 1982; Kareskay et al. 1977; Rowden et al. 1977; Stingl et al. 1978; Swift et al. 2001). Between early and late adulthood, there is a 20–50% decrease in the number of Langerhans cells in the epidermis, and the remaining cells display morphologic abnormalities (Sauder 1986). A reduction in antigen-presenting cells has also been reported in aged (compared with young) mice, with a reduction in both lymphocyte toxicity and a reduced number of $CD8^+$ T cells (Donnini et al. 2002).

A decline in the number and function of sebaceous glands with advancing age causes a decrease in sebum secretion (Engelke et al. 1997; Pochi et al. 1979). Sebum is a complex group of oils that function to protect the skin against friction and make it more impervious to moisture. With advancing age, the size of the remaining sebaceous glands also tends to decrease (Zouboulis and Boschnakow 2001). The age-dependent reduction in sebum secretion can subsequently result in the drying and cracking of the skin (Beauregard and Gilchrest 1987; Makrantonaki and Zouboulis 2007), facilitating the subsequent entry of pathogens through the broken skin. Collectively, the compromised integrity of the skin barrier, decrease in the number of macrophages, and reduced ability for activation of the cell-mediated immune function result in an increase in the susceptibility to skin infection and subsequent entry of pathogens into the systemic circulation (Fenske and Lober 1986).

Environmental chemicals that can enter the body through the unbroken skin may thus enter aged skin more easily, and substances that degrade the skin can further challenge the body's first line of defense in protecting against both xenobiotics and pathogens of biologic origin. Environmental substances that can further compromise the integrity of aged skin include those that cause cutaneous burns. Those include ammonia, calcium oxide, chlorine, ethylene oxide, hydrogen chloride, hydrogen fluoride, hydrogen peroxide, methyl bromide, oxides of nitrogen, phosphorous, phenol, sodium hydroxide, and toluene diisocyanate (Klaassen 2001). Cigarette smoking is known to exacerbate the aging of the skin, especially in women (Davis and Koh 1992; Lange and Schnohr 2007; Smith and Fenske 1996).

Contact with some toxicants not only presents a risk of damage to the integument, but also represents a potential route of entry into the body. Hence, environmental substances (e.g., organophosphorus and carbamate insecticides) that are rapidly and effectively absorbed through the intact/unbroken skin of healthy young adults present a potentially greater risk to the elderly, whose compromised skin may increase the rate and extent of dermal absorption (ATSDR 2008, 1993). Similarly, dermal exposure to solvents and petroleum products (e.g., gasoline, kerosene) may result in increased integumentary damage and absorption into the systemic circulation in the elderly, compared to younger individuals.

## 2.10 Aging and the Respiratory System

As with the other organ systems of the human body, the respiratory system undergoes both anatomic and functional decrements with advancing age. The static elastic recoil of the lung decreases with age, making the chest wall easier to expand. One might initially expect this to enable the elderly individual to take deeper breaths; however, the chest wall itself becomes stiffer with aging, and this increasing rigidity has a measurable impact on the mechanics of breathing (Mahler et al. 1986; Mittman et al. 1965). This stiffness causes a reduction in both maximal inspiratory volume and vital capacity (total lung volume minus the reserve volume) (Enright et al. 1993; Hazzard et al. 1999). Both slow and forced vital capacity decline with age. The rate of decline accelerates as age progresses. This decline has been estimated in cross-sectional studies to range from 21 to 34 mL/year in men and from 19 to 29 mL/year in women (Ware et al. 1990).

The respiratory muscles also decrease in strength with aging (Enright et al. 1994; Tolep et al. 1995; Tolep and Kelsen 1993). Chen and Kuo (1989) reported age-related decrements in respiratory muscle strength and endurance of approximately 20% by age 70. Diaphragm muscle strength decreases approximately 25% in healthy elderly persons, compared to the diaphragm strength in young adults, contributing to the decrease in vital capacity.

Pulmonary artery resistance increases significantly with age (Ehrsam et al. 1983). In a population of 1,413 adult subjects (mean age 63 $\pm$ 11 years) with measurable pulmonary artery systolic pressure (PASP), PASP was found to increase with age (Lam et al. 2009). The diffusing capacity of the lungs declines after age 49 and continues to decline at a rate of about 5% per decade (Hazzard et al. 1999; Oskvig 1999). This decrease in gas exchange correlates with the decrease in internal surface area of the lung that occurs with aging (Oskvig 1999). The bottom line is that the aging process results in a decrease in the ability of the lungs to function optimally in maintaining homeostasis. Thus, the ability to supply oxygen to the cells of the body and the ability to eliminate carbon dioxide and maintain blood pH, particularly during periods of activity such as exercise, is compromised. The elderly have a significantly reduced response to hypoxia and hypercapnia, attributable to reduced tidal volume (Kronenberg and Drage 1973). Further, the older the individual, the smaller and more delayed the physiologic response to those states will be (Oskvig 1999). The elderly also have a substantially decreased response for vocal cord closure, which markedly increases the risk of aspiration and consequent airway reaction and pulmonary injury (Pontoppidan and Beecher 1960).

Priox et al. (2000) studied nine elderly males (mean age 68.1 $\pm$ 4.8 years) and nine young males (mean age 23.4 $\pm$ 1.3 years) during incremental exercise. The elderly subjects were found to have significantly higher values for minute ventilation, respiratory equivalents for oxygen intake and carbon dioxide output, mean inspiratory flow, and lactate concentration than the young subjects. The authors attribute the increases in ventilatory parameters, in part, to the increased lactate concentrations.

In a study of 34 older subjects (ages 60–75) and 10 young subjects (ages 24–33), Chisari et al. (2002) looked at blood lactate levels before and after incremental exercise on a treadmill. While resting lactate levels before exercise were comparable in both the young and older groups, older subjects showed significantly higher lactate levels during the post-exercise recovery period. They concluded that the abnormal lactate increase seen following exercise in the older group indicated a reduced oxidative muscle function in older people (Chisari et al. 2002).

Exposure to environmental particulates, smoke, or other substances (by any route of exposure) that either cause bronchial constriction or mucus production or that cause circulatory acidosis might be expected to produce more severe results in the elderly individual. The effects would be even more pronounced in an individual with chronic respiratory disease, such as asthma, chronic obstructive pulmonary disease, or emphysema. An impaired beta receptor-mediated bronchodilator response has also been reported in unhealthy elderly individuals, but not among elderly who were healthy (Kradjan et al. 1992).

Medications can further compromise barely adequate respiratory muscle strength and endurance. The elderly also have a much higher incidence of apnea and periodic breathing with narcotics and respiratory depression from benzodiazepines (Oskvig 1999).

Thus, respiratory irritants, such as sulfur dioxide and formaldehyde, may be expected to have a significant effect on the efficacy of the respiratory system of an elderly individual, as compared to the respiratory system of younger, healthy individuals.

## 3 Pharmacology and Chemical/Drug Interactions

The human aging process is linked mechanistically to altered drug handling, altered physiological reserve, and consequent pharmacodynamic responses (McLean and Le Couteur 2004). With aging, the metabolism and excretion of many drugs decrease (Beers et al. 2006). Overall metabolic capacity decreases with advancing age, probably as a result of reduced liver volume and diminished hepatic blood flow (Turnheim 1998). This can have a number of effects on individual drugs, including a decrease in bioavailability of some concurrently administered medications (Dilger et al. 2000).

With the increased use of prescription and over-the-counter (OTC) medications, the potential for drug–toxicant interaction also increases (Allard et al. 2001). And since the elderly take more drugs than their young counterparts, they are more susceptible to adverse drug interactions (Allard et al. 2001; Parker et al. 1995; Reidenberg 1982). Adverse drug events (ADEs) increase with greater age (Ganjavi et al. 2007). ADEs are the most common type of adverse event in hospitalized patients, including those of age 65 years or age or older and such events are also common in nursing homes (Leape et al. 1991; Rothschild et al. 2000).

At present, the elderly comprise approximately 12% of the US population, but they consume approximately 25% of the prescription drugs sold annually. In addition to prescription drugs, the elderly frequently use OTC medications to treat their symptoms or illnesses, and they often rely on multiple physicians, including specialists, to diagnose their illnesses and provide relief from pain and other undesirable symptomatology (Qato et al. 2008). One in three older adults in the United States uses five or more prescription medications regularly (Qato et al. 2008). In a recent study of elderly Germans, it was found that older general practice patients consumed a mean of 3.7 prescribed medications and an additional 1.4 OTC drugs (Junius-Walker et al. 2006). Hui-Ling et al. (2008) reported that the mean number of medications per long-term nursing care resident in Taiwan was $5.74 \pm 2.4$. Of these 25.1% had experienced a drug–drug interaction, 64.95% of which were moderate in severity and 7.2% were of major severity. In the period 2005–2006, at least 1 in 25 older US adults used a drug regimen posing a risk of major potential drug–drug interaction, half of which involved the use of non-prescription medications (Qato et al. 2008). Wu (2000) reviewed some of the studies regarding adverse drug events, many of which occurred in the elderly. In that study, as well as others, it was pointed out that older patients are highly vulnerable to the adverse effects of drugs (Einerson 1993; Hamilton et al. 2009; Hanlon et al. 1997; Williamson and Chopin 1980). Contributors to this phenomenon included greater physiologic susceptibility among the elderly, along with less functional reserve, problems with recall, care from multiple physicians, and the use of more than one pharmacy (Col et al. 1990). The most commonly used drugs associated with excessive polypharmacy in the Kuopio 75+ Finish study were cardiovascular drugs and analgesics (Jyrkka et al. 2009).

The high prevalence of polypharmacy in the elderly likely contributes to an abnormally high incidence (20–25%) of adverse drug reactions in this age group (Hunt et al. 1992a, b; McClean and Le Couteur 2004). Oliver et al. (2009) studied patients aged 65 and older admitted to a French hospital and reported that a significant incidence of adverse drug reactions leading to hospitalization was found among elderly patients. The most important factors leading to this were the number of drugs being taken, self-medication, the use of anti-thrombotics, and the use of anti-bacterial drugs (Oliver et al. 2009). Sharkey et al. (2005) looked at the patterns of therapeutic prescription medication use by category among community-dwelling homebound older adults and reported that more than 40% of the individuals studied took medications from three to four different therapeutic categories.

Budnitz et al. (2007) reported an estimated 175,000 emergency department visits annually for adverse drug events, one-third of which were attributed to adults over 65 years of age. When combined with exposure to environmental chemicals, including those commonly used in homes, the potential for chemical (including pharmaceuticals)-to-chemical reactions is significant.

In a study conducted in Finland, Linjakumpu et al. (2002) reported that polypharmacy among the elderly was on the increase. The number of medications per person had risen between the 1990–1991 and 1998–1999 sampling periods from 3.1 to 3.8, and the concomitant use of more than 5 medications has increased from 19 to 25% during the same period. The most commonly used medications were cardiovascular and CNS drugs (Linjakumpu et al. 2002). More recently Qato et al. (2008) have reported the increased use of dietary supplements concomitant with prescription medication and OTC drugs among older adults.

In a study of hospital admissions, Juurlink et al. (2003) found that elderly patients admitted for drug toxicity were exposed to drugs known to cause drug–drug interactions. Data obtained from institutionalized elderly patients showed a mean daily intake of three to eight drugs, with a somewhat higher use of psychotrophic drugs compared with non-institutionalized elderly living in communities. This has resulted in adverse sequelae related to the number of drugs used. In some instances, pharmacodynamic drug interactions can even result in the alteration of the physiologic response to one drug without altering the concentration of that drug (Hazzard et al. 1999; Monette et al. 1995; Walker and Wynne 1994). Junius-Walker et al. (2006) reported that 26.7% of 466 patients aged 70+ years used five or more chronically prescribed drugs, resulting in health effects including breathlessness, hypertension, and low subjective health.

The use of pharmaceuticals known to be contraindicated for elderly patients creates yet another source of risk for the elderly. Wilcox et al. (1994) examined the records of 6,171 people aged 65 or older living in elderly communities. Of 20 drugs contraindicated for the elderly, they found that 23.5% (equivalent to 6.64 million Americans overall) received at least one of the contraindicated drugs. Further, 20.4% of the 23.5% received two or more contraindicated drugs. The authors concluded that physicians prescribe potentially inappropriate medications for nearly a quarter of all older people living in such communities, placing them at risk of adverse effects, such as cognitive impairment and sedation (Wilcox et al. 1994). In a related study, it was reported that in the year 1996, 21.3% of community-dwelling elderly patients in the United States received at least 1 of 33 potentially inappropriate medications, and about 2.6% of the elderly adults studied used at least 1 of 11 medications that should always be avoided by elderly patients (Zhan et al. 2001).

Drugs used to control urinary incontinence are widely used among the elderly. Ruby et al. (2005) reported that 9.5% of men and 54.0% of women studied had difficulty holding urine. While anti-cholinergics are often prescribed for this purpose, other drugs commonly used among the elderly can have effects that indirectly antagonize these drugs. Diuretics used to treat hypertension, common among the elderly, cause polyuria, increasing the risk of urinary control problems. Benzodiazepines and anti-depressants, two classes of drugs widely used in elderly populations, affect sensory neurologic input, and thus cognition and mobility, two reported risk factors for urinary incontinence (Steele et al. 1999). Anti-psychotics used to treat age-related dementia can have similar effects. Beta blockers and alpha-adrenergic antagonists, such as prazosin and clonidine, increase bladder contractility and decrease outlet resistance, thus increasing the risk of urinary incontinence (Ruby et al. 2005). Acetylcholinesterase inhibitors used to treat Alzheimer's disease (Aricept, Razadyne, and Exelon) can similarly have an adverse effect on one's ability to control their bladder, as can environmental cholinesterase inhibitors such as organophosphorus and carbamate insecticides.

Further, the body's fat compartment increases with aging, increasing the distribution for highly lipophilic drugs and increasing their elimination half-lives. As a result, levels of some chronically used drugs tend to increase for about six half-lives, leading to a gradual build-up to toxic levels (Beers et al. 2006).

The use of prescription and over-the-counter medications, which is typical among the aged, may also affect the manner in which the body deals with environmental toxicants. For example, a person taking ibuprofen regularly for pain management or joint inflammation will be more susceptible to the toxic effects of environmental substances that affect the liver or kidney.

Barrett (2009) reported that an estimated 41 million Americans are exposed to trace pharmaceuticals in their drinking water, according to the results of an Associated Press Investigation published in March of 2008. Even at very low concentrations, this may present an additional environmental challenge to elderly adults who have an already-compromised metabolic capability and are taking a variety of pharmaceutical medications. Xenobiotics having neurologic, cardiovascular, renal, or immunologic effects may present additional physiologic and homeostatic challenge to this age-weakened and metabolically challenged population.

## 4 Are Existing Health Guidance Values Adequately Protective of a Compromised Population?

In the calculation of health guidance values (HGVs), such as the Agency for Toxic Substances and Disease Registry's (ATSDR's) Minimal Risk Levels (MRLs) and the US Environmental Protection Agency's (EPA's) oral Reference Doses (RfDs) and inhalation Reference Concentrations (RfCs), an uncertainty factor of 1–10 is routinely applied to afford protection to the most sensitive individuals within the human population (Barnes and Dourson 1988; Chou et al. 1998). Infants, small children, and pregnant women are typically considered to be those most sensitive to xenobiotics, because the processes involved in human development are on-going in those individuals. In those instances in which HGVs are based upon data obtained from a study population consisting of such individuals, an uncertainty factor of 3 or 10 is sometimes used, depending on a variety of factors (Chou et al. 1998; Risher and De Rosa 1997).

The elderly undergo unique, yet predictable, physiological and anatomical changes that can impact the way their bodies respond to environmental challenges. Since these changes are well known and can be expected to occur in all aging populations, regardless of ethnic or racial origin, the potential for increased susceptibility should be considered when evaluating the health risk of all chemical agents, whether naturally occurring or of anthropogenic origin. When an exposed population is known to include senior citizens, and when exposure to the substance under investigation is known to affect an organ or organ system likely to be compromised by the aging process, the evaluation of the potential health risk should include careful consideration of the compromised physiological state of that population. Public health officials should be aware of the particular problems of the elderly and not merely assume that they are no more vulnerable to environmental toxicants than are the very young.

This chapter has addressed how anatomical, physiologic, nutritional, medical, and behavioral changes in late stages of life may compromise the ability of the elderly to homeostatically deal with exposures to many types of chemicals. The question thus arises as to whether existing health guidance values, such as ATSDR's MRLs, adequately protect the health of the elderly. Risher and De Rosa (1997) state that the consideration of when and how to apply any HGV must be viewed in light of the exposed population. For example, when an MRL is based upon a healthy juvenile or adult population and an uncertainty factor of less than 10 has been employed for the study population then that MRL might not *necessarily* be fully protective of an elderly population in which exposure is likely. This does not mean, however, that the existing HGVs are inadequate; rather, such values must be viewed by public health officials with full consideration given to the exposed population. In the case of the elderly, this might mean that consideration should be given to downward adjustment of the HGV to afford adequate protection if the study population on which that HGV is based does not include elderly subjects.

# 5 Summary

The US population is aging. CDC has estimated that 20% of all Americans will be 65 or older by the year 2030. As a part of the aging process, the body gradually deteriorates and physiologic and metabolic limitations arise. Changes that occur in organ anatomy and function present challenges for dealing with environmental stressors of all kinds, ranging from temperature regulation to drug metabolism and excretion. The elderly are not just older adults, but rather are individuals with unique challenges and different medical needs than younger adults. The ability of the body to respond to physiological challenge presented by environmental chemicals is dependent upon the health of the organ systems that eliminate those substances from the body. Any compromise in the function of those organ systems may result in a decrease in the body's ability to protect itself from the adverse effects of xenobiotics. To investigate this issue, we performed an organ system-by-organ system review of the effects of human aging and the implications for such aging on susceptibility to drugs and xenobiotics.

Birnbaum (1991) reported almost 20 years ago that it was clear that the pharmacokinetic behavior of environmental chemicals is, in many cases, altered during aging. Yet, to date, there is a paucity of data regarding recorded effects of environmental chemicals on elderly individuals. As a result, we have to rely on what is known about the effects of aging and the existing data regarding the metabolism, excretion, and adverse effects of prescription medications in that population to determine whether the elderly might be at greater risk when exposed to environmental substances. With increasing life expectancy, more and more people will confront the problems associated with advancing years. *Moreover, although proper diet and exercise may lessen the immediate severity of some aspects of aging, the process will continue to gradually degrade the ability to cope with a variety of injuries and diseases.* Thus, the adverse effects of long-term, low-level exposure to environmental substances will have a longer time to be manifested in a physiologically weakened elderly population. When such exposures are coupled with concurrent exposure to prescription medications, the effects could be devastating. Public health officials must be knowledgeable about the sensitivity of the growing elderly population, and ensure that the use of health guidance values (HGVs) for environmental contaminants and other substances give consideration to this physiologically compromised segment of the population.

## References

Abate N, Haffner SM, Garg A, Peshock RM, Grundy SM (2002) Sex steroid hormones, upper body obesity, and insulin resistance. J Clin Endocrinol Metab 87(10):4522–4527

Abe O, Yamasue H, Aoki S, Suga M, Yamada H, Kasai K, Masutani Y, Kato N, Kato N, Ohtomo K (2008) Aging in the CNS: comparison of gray/white matter volume and diffusion tensor data. Neurobiol Aging 29:102–116

Abou-Donia MB, Wilmarth KR, Abdel-Rahman AA, Jensen KF, Oehme FW, Kurt TL (1996) Increased neurotoxicity following concurrent exposure to pyridostigmine bromide, DEET, and chlorpyrifos. Fund Appl Toxicol 34:201–222

Ackermann MF, Gasiewicz TA, Lamm KR, Germolec DR, Luster MI (1989) Selective inhibition of polymorphonuclear neutrophil activity by 2,3,7,8-tetrachlorodibenzo-*p*-doixin. Toxicol Appl Pharmacol 101:470–480

ACOG (2004) American college of obstetricians and gynecologists. Task force on hormone therapy. Obstet Gynecol 104(S4):112S

Actis GC, Debernardo-Venon W, Lagget M, Marzano A, Ottobrelli A, Ponzetto A, Rocca G, Boggio-Bertinet D, Balzola F, Bonino F, Verme G (1995) Hepatotoxicity of intravenous cyclosporin A in patients with acute ulcerative colitis on total parenteral nutrition. Liver 15(6):320–323

Agrawal AK, Shapiro B (2003) Constitutive and inducible hepatic cytochrome P450 isoforms in senescent male and female rats and response to low-dose phenobarbital. Drug Metab Dispos 31:612–619

Agrawal Y, Platz EA, Niparko JK (2008) Prevalence of hearing loss and differences by demographic characteristics among US adults. Arch Intern Med 168(14):1522–1530

Albright JW, Albright JF (1998) Impaired natural killer cell function as a consequence of aging. Exp Gerontol 33:13–25

Allard J, Hebert R, Rioux M, Asselin J, Voyer L (2001) Efficacy of a clinical medication review on the number of potentially inappropriate prescriptions prescribed for community-dwelling elderly people. Can Med Assoc J 164(9):1291–1296

Allen JS, Bruss J, Brown CK, Damasio H (2005) Normal neuroanatomical variation due to age: the major lobes and a parcellation of the temporal region. Neurobiol Aging 26:1245–1260

Allen JR, Barsotti DA, Van Miller JP, Abrahamson LJ, Lalich JJ (1977) Morphological changes in monkeys consuming a diet containing low levels of 2,3,7,8-tetrachlorodibenzo-*p*-dioxin. Food Cosmet Toxicol 15:401–410

Allen JR, Carstens LA, Barsotti DA (1974) Residual effects of short-term, low-level exposure of nonhuman primates to polychlorinated biphenyls. Toxicol Appl Pharmacol 30:440–451

Allis JW, Ward TR, Seely JC, Simmons JE (1990) Assessment of hepatic indicators of subchronic carbon tetrachloride injury and recovery in rats. Fundam Appl Toxicol 15:558–570

Aly NM, El-Gendy KS (2000) Effect of dimethoate on the immune system of female mice. J Environ Sci Health Part B 35(1):77–86
Anantharaju A, Feller A, Chedid A (2002) Aging liver: a review. Gerontology 48(6): 343–353
Anderson S, Brenner BM (1986) Effects of aging on the renal glomerulus. Am J Med 80(3): 435–442
Andrews JE (1989) Polychlorinated biphenyl (Aroclor 1254) induced changes in femur morphometry calcium metabolism and nephrotoxicity. Toxicology 57:83–96
Aoyagi M, Fukai N, Ogami K, Yamamoto M, Yamamoto K (1995) Kinetics of 125I-PDGF binding and down-regulation of PDGF receptor in human arterial smooth muscle cell strains during cellular senescence in vitro. J Cell Physiol 164:376–384
Aoyama K (1986) Effects of benzene inhalation on lymphocyte subpopulations and immune response in mice. Toxicol Appl Pharmacol 85:92–101
Aramaki T, Katsuta Y, Sekiyama T, Tsutsui H, Kome-ichi H, Ohsuga M, Satomura K, Yoshimura M (1998) Effects of aging and liver disease upon the pharmacokinetics of nipradilol. Clin Drug Invest 16(3):251–257
Ashcroft GS, Horan MA, Ferguson MW (1997) The effects of aging on wound healing: immunolocalisation of growth factor and their receptors in a murine incisional model. J Anat 190:361–365
Ashcroft GS, Horal MA, Ferguson MWJ (1995) The effects of ageing on cutaneous wound healing in mammals. J Anat 187:1–26
ATSDR (1999) Toxicological profile for mercury (Update). Department of Health and Human Services, Atlanta, GA
ATSDR (1993) Case studies in environmental medicine: cholinesterase-inhibiting pesticide toxicity. Department of Health and Human Services, Atlanta, GA
Bach B, Molholm-Hansen J, Kampriann JP, Rasmussen SN, Skorsted Z (1981) Disposition of antipyrine and phenytoin correlated with age and liver volume in man. Clin Pharmacokinet 6:389–396
Bagai A, Thavendiranathan P, Detsky AS (2006) Does this patient have hearing impairment? JAMA 295(4):416–428
Bailey AJ, Knott L (1999) Molecular changes in bone collagen in osteoporosis and osteoarthritis in the elderly. Exper Gerontol 34(3):337–351
Balin AK, Allen RG (1989) Molecular bases of biologic aging. Clin Geriatr Med 5:1–21
Banks WA, Kastin AJ (1989) Aluminum-induced toxicity: Alterations in membrane function at the blood–brain barrier. Neurosci Biobehav Rev 13:47–53
Banks WA, Farr SA, Morley JE (2000) Permeability of the blood–brain barrier to albumin and insulin in the young and aged SAMP8 mouse. J Gerontol Series A: Biol Sci Med Sci 55: B601–B606
Barnes DG, Dourson M (1988) Reference dose (RfD): description and use in health risk assessments. Regul Toxicol Pharmacol 8:471–486
Barnes R, Jones RC (1967) Carbon tetrachloride poisoning. Am Ind Hyg Assoc J 28:557–560
Barrett JR (2009) Aging. Environmental threats to elders' neurologic health. Environ Health Perspect 117(1):A17
Barsotti DA, Marlar RJ, Allen JR (1976) Reproductive dysfunction in Rhesus monkeys exposed to low levels of polychlorinated biphenyls (Aroclor 1248). Food Cosmet Toxicol 14: 99–103
Bauer JH (1993) Age-related changes in the renin–aldosterone system. Physiological effects and clinical implications. Drugs Aging 3(3):238–245
Baylis C (2005) Changes in renal hemodynamics and structure in the aging kidney; sexual dimorphism and the nitric oxide system. Exp Gerontol 40(4):271–278

Beauregard S, Gilchrest BA (1987) A survey of skin problems and skin care regimens in the elderly. Arch Dermatol 123(12):1638–1643

Beers MH, Porter RS, Jones TV, Kaplan JL, Berkwits M (eds) (2006) The Merck manual, 18th edn. Merck Research Laboratories, Whitehouse Station, NJ

Beers MH, Berkow R (eds) (2000) Merck manual of geriatrics, 10th edn. Merck Research Laboratories, Whitehouse Station, NJ

Begin ME, Langlois MF, Lorrain D, Cunnane SC (2008) Thyroid function and cognition during aging. Curr Gerontol Geriatr Res. doi:10.1155/2008/474868

Belmin J, Levy BI, Michel JB (1994) Changes in the renin–angiotensin–aldosterone axis in later life. Drugs Aging 5(5):391–400

Benetos A, Laurent S, Hoeks AP, Boutouyrie PH, Safar ME (1993) Arterial alterations with aging and high blood pressure: a non-invasive study of carotid and femoral arteries. Atheroscler Thromb: 13:90–97

Benjamin IL, Jalil JE, Tan L-B, Cho K, Weber KT, Clark WA (1989) Isoproterenol-induced myocardial fibrosis in relation to myocyte necrosis. Circ Res 65:657–670

Ben-Yehuda A, Weksler ME (1992) Host resistance and the immune system. Clin Geriatr Med 8:701–711

Beyer WE, Palache AM, Baljet M, Masurel N (1989) Antibody induction by influenza vaccines in the elderly: a review of the literature. Vaccine 7(5):385–394

Bianchi L, Holt P, James OFW, Butler RN (eds) (1988) Aging in liver and gastrointestinal tract. Kluwer, Hingham, MA

Binkley N, Novotny R, Krueger D, Kawahara T, Daida YG, Lensmeyer G, Hollis BW, Drezner MK (2007) Low vitamin D status despite abundant sun exposure. J Clin Endocrinol Metab 92(6):2130–2135

Birnbaum LS (1991) Pharmacokinetic basis of age-related changes in sensitivity to toxicants. Ann Rev Pharmacol 31:101–128

Blair PC, Thompson MB, Wilson RE, Esber HH, Maronpot RR (1991) Correlation of changes in serum analytes and hepatic histopathology in rats exposed to carbon tetrachloride. Toxicol Lett 55:149–159

Blakley BR (1985) The effect of cadmium chloride on the immune response in mice. Can J Comp Med 49:104–108

Blaylock BL, Holladay SD, Comment CE, HeGindel J, Luster MI (1992) Exposure to tetrachlorodibenzo-*p*-dioxin (TCDD) alters fetal thymocyte maturation. Toxicol Appl Pharmacol 112:207–213

Bogardus ST, Yueh B, Shekelle PG (2003) Screening and management of adult hearing loss in primary care: clinical applications. JAMA 289(15):1986–1990

Bonnick SL (1994) The osteoporosis handbook. Taylor, Dallas, TX

Born J, Uthgenannt D, Dodt C, Ninninghoff D, Ringvolt E, Wagner T, Fehm HL (1995) Cytokine production and lymphocyte subpopulations in aged humans. An assessment during nocturnal sleep. Mech Ageing Dev 84:113–126

Boyd E (1932) The weight of the thymus gland in health and disease. Am J Dis Child 43: 1162–1214 (as cited in Francis et al. 1985)

Brace H, Latimer M, Winn P (1997) Neurotoxicity, blood–brain barrier breakdown, demyelination and remyelination associated with NMDA-induced lesions of the rat lateral hypothalamus. Brain Res Bull 43:447–455

Branca F (1997) Calcium, micronutrients and physical activity to maximize bone mass during growth. Food Nutr Agric 20:44–48

Branchet MC, Boisnic S, Frances C, Robert AM (1990) Skin thickness changes in normal aging. Gerontology 36(1):28–35

Braverman IM (2000) The cutaneous microcirculation. J Invest Dermatol Symp Proc 5: 3–9

Braverman IM, Fonferko E (1982) Studies in cutaneous aging. II. The microvasculature. J Invest Dermatol 78:444–448

Bruckner JV, MacKenzie WF, Muralidhara S, Luthra R, Kyle GM, Acosta D (1986) Oral toxicity of carbon tetrachloride: acute, subacute and subchronic studies in rats. Fundam Appl Toxicol 6:16–34

Bruckner JV, Jiang WD, Brown JM, Putcha CK, Chuand VJ, Stella J (1977) The influence of ingestion of environmentally encountered levels of a commercial polychlorinated biphenyl mixture (Aroclor 1254) on drug metabolism in the rat. J Pharmacol Exp Ther 202:22–31

Bruckner JV, Khanna KL, Cornish HH (1973) Biological responses of the rat to polychlorinated biphenyls. Toxicol Appl Pharmacol 24:434–448

Budnitz DS, Shehab N, Kegler SR, Richards CL (2007) Medication use leading to emergency department visits for adverse drug effects in older adults. Ann Intern Med 147:755–765

Burger H, de Laet CEDH, van Daele PLA, Weel AEAM, Witterman JCM, Hofman A, Polsw HAP (1998) Risk factors for increased bone lows in an elderly population: the Rotterdam study. Am J Epidemiol 147:871–879

Buschbacher RM, Koch J (1999) Race effect on nerve conduction studies: Acomparison between 50 blacks and 50 whites. Arch Phys Med Rehabil 80:536–539

Butcher SK, Chahal H, Nayak L, Sinclair A, Henriques NV, Sapey E, O'Mahony D, Lord JM (2001) Senescence in innate immune responses: reduced neutrophil phagocytic capacity and CD16 expression in elderly humans. J Leukoc Biol 70:881–886

Campbell VA, Crews JE, Moriarty DG, Zack MM, Blackman DK (1999) Surveillance for sensory impairment, activity limitation, and health-related quality of life among older adults – United States, 1993–1997. Morb Mortal Wkly Rep 48:131–156

Carter JW (1985) Effects of dietary PCBs (Aroclor 1254) on serum levels of lipoprotein cholesterol in Fischer rats. Bull Environ Contam Toxicol 34:427–431

Carter JW, Koo SI (1984) Effects of dietary Aroclor 1254 (PCBs) on serum levels of lipoprotein cholesterol and tissue distribution of zinc, copper and calcium in Fischer rats. Nutr Rep Int 29:223–232

Castle S, Uyemura K, Wong W, Modlin R, Efros R (1997) Evidence of enhanced type 2 immune response and impaired upregulation of a type 1 response in frail elderly nursing home residents. Mech Ageing Dev 94:7–16

CDC (2004) The state of aging and health in America 2004. Merck Institute of Aging & Health, Washington, DC

Cha CW (1987) A study on the effect of garlic to the heavy metal poisoning of rat. J Korean Med Sci 2:213–224

Chahal HS, Drake WM (2007) The endocrine system and aging. J Pathol 211(2):173–180

Chakraborti AK, Saha KC (1987) Arsenical dermatosis from tubewell water in West Bengal. Indian J Med Res 85:326–334

Chandra AM, Campbell GA, Reddy G, Qualls CW (1999) Neurotoxicity of 1,3,5-trinitrobenzene (TNB): immunohistochemical study of cerebrovascular permeability. Vet Pathol 36:212–220

Chang E, Yang J, Nagavarapu U, Herron GS (2002) Aging and survival of cutaneous microvasculature. J Invest Dermatol 118:7542–7758

Chen C-H, Nakayama M, Nevo E, Fetics BJ, Maughan WL, Kass DA (1998) Coupled systolic-ventricular and vascular stiffening with age. J Am Coll Cardiol 32:1221–1227

Chen HS, Kuo CS (1989) Relationship between respiratory muscle function and age, sex, and other factors. J Appl Physiol 66:943–948

Chertkov JL, Lutton JD, Jiang S, da Silva JL, Abraham NG (1992) Hematopoietic effects of benzene inhalation assessed by murine long-term bone marrow culture. J Clin Lab Med 119:412–419

Chisari C, Bresci M, Licitra R, Stampacchia G, Rossi B (2002) A functional study of oxidative muscle efficiency in older people. Basic Appl Myol 12(5):209–212

Chou SJ, Holler J, De Rosa CT (1998) Minimal risk levels for hazardous substances. J Clean Technol Environ Toxicol Occup Med 7(1):1–24

Chutka DS, Flemming KC, Evans MP, Evans JM, Andrews KL (1996) Urinary incontinence in the elderly population. Mayo Clin Proc 71:93–101

Clewell RA, Merrill EA, Narayanan L, Gearhart JM, Robinson PJ (2004) Evidence for competitive inhibition of iodide uptake by perchlorate and translocation of perchlorate in to the thyroid. Int J Toxicol 23(1):17–23

Cody RP, Strawderman WW, Kipen HM (1993) Hematologic effects of benzene. Job-specific trends during the first year of employment among a cohort of benzene-exposed rubber workers. J Occup Med 35(8):776–782

Col N, Fanale JE, Kronholm P (1990) The role of medication noncompliance and adverse drug reactions in hospitalizations of the elderly. Arch Intern Med 150:841–845

Colige A, Nusgens B, Lapiere CM (1990) Response to epidermal growth factor of skin fibroblasts from donors of varying age is modulated by the extracellular matrix. J Cell Physiol 145: 450–457

Colonna-Romano G, Aquino A, Bulati M, Di Lorenzo G, Listi F, Vitello S, Lio D, Candore G, Clesi G, Caruso C (2006) Memory B cell subpopulations in the aged. Rejuvenation Res 9: 149–152

Condie LW, Laurie RD, Mills T, Robinson M, Bercz JP (1986) Effect of gavage vehicle on hepatotoxicity of carbon tetrachloride in CD-1 mice: corn oil versus Tween-60 aqueous emulsion. Fundam Appl Toxicol 7:199–206

Cooke GA, AL-Timman JK, Marshal P, Wright DJ, Hainsworth R, TanL-B (1998) Physiological cardiac reserve: development of a non-invasive method and first estimates in man. Heart 79:289–294

Corpas E, Harman SM, Blackburn MR (1993) Human growth hormone and human aging. Endocr Rev 14:20–39

Costa-Bauza A, Ramis M, Montesinos V, Grases F, Conte A, Piza P, Pieras E (2007) Type of renal calculi: variation with age and sex. World J Urol 25(4):415–421

Courchesne E, Chisum HJ, Townsend J, Cowles A, Covington J, Egaas B, Harwood M, Hinds S, Press GA (2000) Normal brain development and aging: quantitative analysis at in vivo MR imaging in healthy volunteers. Radiology 216(3):672–682

Cronkite EP, Drew RT, Inoue T, Bullis JE (1989) Hematotoxicity and carcinogenicity of inhaled benzene. Environ Health Perspect 82:97–198

Cronkite EP (1986) Benzene hematotoxicity and leukemogenesis. Blood Cells 12:129–137

Cronkite EP, Drew RT, InoueT, Bullis JE (1985) Benzene hematotoxicity and leukemogenesis. Am J Ind Med 7:447–456

Crocenzi FA, Pellegrino JM, Catania VA, Luquita MG, Roma MG, Mottino AD, Sanchez Pozzi EJ (2006) Galactosamine prevents ethinylestradiol-induced cholestasis. Drug Metab Disp DOI: 10.1124/dmd.106.009308

David A, Frantik E, Holusa R, Novakova O (1981) Role of time and concentration on carbon tetrachloride in rats. Int Arch Occup Environ Health 48:49–60

Davis BE, Koh HK (1992) Faces going up in smoke: a dermatologic opportunity for cancer prevention. Arch Dermatol 128:1106–1107

De Heer C, Verlaan APJ, Penninks AH, Vos JG, Schuurman HJ, Van Loveren H (1994) Time course of 2,3,7,8-tetrachlorodibenzo-*p*-dioxin (TCDD)-induced thymic atrophy in the Wistar rat. Toxicol Appl Pharmacol 128:97–104

De Krey GK, Kerkvliet NI (1995) Suppression of cytotoxic T lymphocyte activity by 2,3,7, 8-tetrachlorodibenzo-*p*-dioxin occurs in vivo, but not in vitro, and is independent of corticosterone elevation. Toxicology 97:105–112

De la Cruz Rodriguez LC, Araujo CR, Posleman SE, Del Rosario Rey M (2007) Hepatotoxic effect of cyclosporin A in the mitochondrial respiratory chain. J Appl Toxicol 27:310–317

De la Fuente M, Minano M, Victor VM, Del Rio M, Ferrabdez MD, Diez A, Miguel J (1998) Relation between exploratory activity and immune function in aged mice: a preliminary study. Mech Aging Dev 102:263–277

De Wall EJ, Schuurman H-J, Loeber JG, Van Loveren H, Vos JG (1992) Alterations in cortical thymic epithelium of rats after in vivo exposure to 2,3,7,8-tetrachlorodibenzo-*p*-dioxin (TCDD): an (immuno)histological study. Toxicol Appl Pharmacol 115(1): 80–88

DeCaprio AP, McMartin DM, O'Keefe PW, Rey R, Silkworth JB, Kaminsky LS (1986) Subchronic oral toxicity of 2,3,7,8-tetrachlorodibenzo-*p*-dioxin in the guinea pig: comparisons with a PCB-containing transformer fluid pyrolysate. Fundam Appl Toxicol 6:454–463

Delarosa O, Pawelec G, Peralbo E, Wikby A, Mariani E, Mocchegiani E, Tarazona R, Solana R (2006) Immunological biomarkers of ageing in man: changes in both innate and adaptive immunity are associated with health and longevity. Biogerontology 7:471–481

Deng Y, Jing Y, Campbell AE, Gravenstein S (2004) Age-related impaired type 1 T cell responses to influenza: reduced activation ex vivo, decreased expansion in CTL culture in vitro, and blunted response to influenza vaccination in vivo in the elderly. J Immunol 172(6):3437–3446

Dennison E, Hindmarsh P, Fall C, Kellingray S, Barker D, Phillips D, Cooper C (1999) Profiles of endogenous circulating cortisol and bone mineral density in healthy elderly men. J Clin Endocrinol Metab 84(9):3058–3063

DeVeale B, Brummel T, Seroude L (2004) Immunity and aging: the enemy within? Aging Cell 2004:195–208

Dieter MP, Boorman GA, Jameson CW, Eustis SL, Uriah LC (1992) Development of renal toxicity in F344 rats gavaged with mercuric-chloride for 2 weeks, or 2,4,6, 15, and 24 months. J Toxicol Environ Health 36(4):319–340

Dieter MP, Luster MI, Boorman GA, Jameson GW, Cox JW (1983) Immunological and biochemical responses in mice treated with mercuric chloride. Toxicol Appl Pharmacol 68:218–228

Dilger K, Hofmann U, Klotz U (2000) Enzyme induction in the elderly: effect of rifampin on the pharmacokinetics and pharmacodynamics of propafenone. Clin Pharmacol Therapeut 67: 512–520

Diliberto JJ, Akubue PI, Luebke RW, Birnbaum LS (1995) Dose–response relationships of tissue distribution and induction of CYP1A1 and CYP1A2 enzymatic activities following acute exposure to 2,3,7,8-tetrachlorodibenzo-*p*-dioxin (TCDD) in mice. Toxicol Appl Pharmacol 130:197–208

Ding A, Hwang S, Schwab R (1994) Effect of aging on murine macrophages – Diminished response to IFN-gamma for enhanced oxidative metabolism. J Immunol 153:2146–2152

Donnini A, Argentati K, Mancini R, Smoriesi A, Bartozzi B, Benardini G, Provinciali M. (2002) Phenotype, antigen-presenting capacity, and migration of antigen-presenting cells in young and old age. Exp Gerontol 37(8–9):1097–1112

Dorfman LJ, Bosley TM (1979) Age-related changes in peripheral and central nerve conduction in man. Neurology 29:38–44

Dow (1992) Effects of benzene vapor in the pig and rat. I. Pertaining to hematology and immunology with cover letter dated 05/14/92. Submitted to the U.S. Environmental Protection Agency under TSCA section 8E. OTS0539784

Edelson GW, Kleerekoper M (1995) Bone mass, bone loss, and fractures. In: Matkovic V (ed) Physical medicine and rehabilitation clinics of North America. WB Saunders, Philadelphia, PA, pp 455–464

Ehrsam RE, Perruchaud A, Oberholzer M, Burkart F, Herzog F (1983) Influence of age on pulmonary hemodynamics at rest and during supine exercise. Clin Sci 65:653–660

Einerson TR (1993) Drug-related hospital admissions. Ann Pharmacother 27:832–840

El-Domyati M, Attia S, Saleh F, Brown D, Birk DE, Gasparro F, Ahmad H, Uitto J (2002) Intrinsic aging vs. photoaging: a comparative histopathological, immunohistochemical, and ultrastructural study of skin. Exp Dermatol 11:398–405

Emeigh Hart SG, Wyand SD, Khairallah EA, Cohen SD (1996) Acetaminophen nephrotoxicity in the CD-1 mouse. II. Protection by probenecid and AT-125 without diminution of renal covalent binding. Toxicol Appl Pharmacol 136:161–169

Emeigh Hart SG, Beierschmitt WP, Wyand DS, Khairallah EA, Cohen SD (1994) Acetaminophen nephrotoxicity in CD-1 mice. I. Evidence of a role for in situ activation in selective covalent binding and toxicity. Toxicol Appl Pharmacol 126:267–275

Emmanuelli G, Lanzio M, Anfossi T, Romano S, Anfossi G, Calamuggi G (1986) Influence of age on polymorphonuclear leukocytes in vitro: phagocytic activity in healthy human subjects. Gerontology 32:308–316

Endress DB, Morgan CH, Garry PJ, Omdahl JL (1987) Age related changes in serum immunoreactive parathyroid hormone and its biological activity in healthy men and women. J Clin Endocrinol Metab 65:724–731

Engelke M, Jensen JM, Ekanayake-Mudiyanselage S, Proksch E (1997) Effects of xerosis and ageing on epidermal proliferation and differentiation. Br J Dermatol 137:219–225

Enright PL, Kronmal RA, Higgins M, Schenker M, Haponik EF (1993) Spirometry M.L. Reference values for women and men 65–85 years of age. Cardiovascular Health Study. Am Rev Resp Dis 147:125–133

Enright PL, Kronmal RA, Manolio TA, Schenker M, Hyatt RE (1994) Respiratory muscle strength in the elderly: correlates and reference values. Am J Respir Crit Care Med 149:430–438

Erdogan MF (2003) Thiocyanate overload and thyroid disease. Biofactors 19(3–4):107–111

Fan F, Wierda D, Rozman KK (1996) Effects of 2,3,7,8-tetrachlorodibenzo-*p*-dioxin on humoral and cell-mediated immunity in Sprague-Dawley rats. Toxicology 106:221–228

Fanelli A, Berlin WK, Grollman EF (1995) Inhibition of iodide transport in rat thyroid cells using *n*-substituted anthranilic acid derivatives. Thyroid 5(3):223–230

Farber BF, Moellering RC Jr (1983) Retrospective study of the toxicity of preparations of vancomycin from 1974 to 1981. Antimicrob Agents Chemother 23:138–141

Farrell SE (2009) Toxicity, acetaminophen: eMedicine in MedScape by WebMD. http://emedicine.medscape.com/article/820200-overview May 24, 2010

Feldman ML (1976) Aging changes the morphology of cortical dendrites. In: Terry RD, Gershon S (eds) Neurobiology of aging. Raven Press, New York, NY, p 11

Fenske NA, Lober CW (1986) Structural and functional changes of normal aging skin. J Am Acad Dermatol 15:571–585

Fickert P, Zatloukal K (2000) Pathogenesis of alcoholic liver disease. In: Zernig G, Saria A, Kurz M, O'Malley S (eds) Handbook of alcoholism. CRC Press, Boca Raton, FL, pp 317–323

Fingerle H, Fischer G, Classen HG (1982) Failure to produce hypertension in rats by chronic exposure to cadmium. Food Chem Toxicol 20:301–306

Forbes GB, Halloran E (1965) The adult decline in lean body mass. Hum Biol 48:162–173

Fournier M, Chevalier G, Nadeau D, Trotter B, Krzystyniak K (1988) Virus–pesticide interactions with murine cellular immunity after sublethal exposure to dieldrin and aminocarb. J Toxicol Environ Health 25:103–118

Francis IR, Gazer GM, Bookstein FL, Gross BH (1985) The thymus: reexamination of age-related changes in size and shape. Am J Roentgen 145(2):249–254

Franzblau A, Lilis R (1989) Acute arsenic intoxication from environmental arsenic exposure. Arch Environ Health 44(6):385–390

Freinkel N, Ingbar SH (1955) Effect of metabolic inhibitors upon iodide transport in sheep thyroid slices. J Clin Endocrinol Metab 15(5):598–615

Ganjavi H, Herrmann N, Rochon PA, Sharma P, Lee M, Cassel D, Freedman M, Black SE, Lanctot KL (2007) Adverse drug events in cognitively impaired elderly patients. Dement Geriatr Cogn Disord 23(6):395–400

Garfinkel S, Hu X, Prudovsky IA, McMahon G, Kapnik EM, McDowell SD, Maciag T (1996a) FGF-1-dependent proliferative and migratory responses are impaired in senescent human umbilical vein endothelial cells and correlate with the inability to signal tyrosine phosphorylation of fibroblast growth factor receptor-1-substrates. J Cell Biol 134:783–791

Garfinkel S, Wessendorf JH, Hu X, Maciag T (1996b) The human diploid fibroblast senescence pathway is independent of interlukin-1 alpha mRNA. Biochim Biophys Acta 1314:109–119

Geokas MC, Lakatta EG, Makinodan T, Timaras PS (1990) The aging process. Ann Intern Med 113:455–466

George DK, Crawford DH (1996) Antibacterial-induced hepatotoxicity. Incidence, prevention and management. Drug Saf 15:79–85

Germain AM, Carvajal JA, Glasinovic JC, Kato S, Williamson C (2002) Intrahepatic cholestasis of pregnancy: an intriguing pregnancy-specific disorder. Gynecol Invest 9:10–14

Geubel AP, De Galocsy C, Alves N, Rahier J, Dive C (1991) Liver damage caused by therapeutic vitamin A administration: estimate of dose-related toxicity in 41 cases. Gastroenterology 100(6):1701–1709

Ghia P, Melchers F, Rolink AG (2000) Age-dependent changes in B lymphocyte development in man and mouse. Exper Gerontol 35:159–165
Ghia P, ten Boekel E, Sanz E, de la Hera A, Rolink A, Melchers F (1996) Ordering of human bone marrow B lymphocyte precursors by single-cell polymerase chain reaction analyses of the rearrangement status of the immunoglobulin H and L chain gene loci. J Exp Med 184: 2217–2229
Ghorbel H, Fetoui H, Mahjoubi A, Guermazi F, Zeghal N (2008) Thiocyanate effects on thyroid function of weaned mice. C R Biol 331(4):262–271
Gilchrest BA, Murphy GF, Soter NA (1982) Effect of chronologic aging and ultraviolet irradiation on Langerhans cells in human epidermis. J Invest Dermatol 79(2):85–88
Gilchrest BA, Blog FB, Szabo G (1979) Effects of aging and chronic sun exposure on melanocytes in human skin. J Invest Dermatol 73:141–143
Gill DP, Jenkins VK, Kempen RR, Ellis S (1980) The importance of pluripotential stem cells in benzene toxicity. Toxicology 16:163–171
Globerson A (1995) T lymphocytes and aging. Int Arch Allergy Immunol 107:491–497
Goldspink DF, Burniston JG, Tan L-B (2003) Cardiomyocyte death and the aging and failing heart. Exp Physiol 88(3):447–458
Goldstein JA, Hickman P, Jue DL (1974) Experimental hepatic porphyria induced by polychlorinated biphenyls. Toxicol Appl Pharmacol 27:437–448
Good CD, Johnsrude IS, Ashburner J, Henson RN, Friston KJ, Frackowiak RS (2001) A voxel-based morphometric study of ageing in 465 normal adult human brains. Neuroimage 14:21–36
Goodwin K, Viboud C, Simonsen L (2006) Antibody response to influenza vaccination in the elderly: a quantitative review. Vaccine 24:1159–1169
Goodwin JS, Messner RP (1979) Sensitivity of lymphocytes to prostaglandin E2 increases in subjects over age 70. J Clin Invest 64(2):434–439
Gopinath B, Rochtchina E, Wang JJ, Schneider J, Leeder SR, Mitchell P (2009) Prevalence of age-related hearing loss in older adults: blue mountains study. Arch Intern Med 169(4):415–416
Gordon HL, Hunter JM (1982) Ethylene glycol poisoning: a case report. Anaesthesia 37(3): 332–338
Goukassian D, Ga F, Yarr M, Eller MS, Nehal US, Gilchrist BA (2000) Mechanisms and implications of the age-associated decrease in DNA repair capacity. FASEB J 14:1325–1334
Gourtsoyiannis N, Prassopouolos P, Cavouras D, Pantelidis N (1990) The thickness of the renal parenchyma decreases with age: a CT study of 360 patients. Am J Roentgen 155:541–544
Graham JA, Miller FJ, Daniels MJ, Payne EA, Gardner DE (1978) Influence of cadmium, nickel, and chromium on primary immunity in mice. Environ Res 16:77–87
Gray LE Jr, Ostby J, Marshall R, Andrews J (1993) Reproductive and thyroid effects of low-level polychlorinated biphenyl (Aroclor 1254) exposure. Fundam Appl Toxicol 20(3):288–294
Gray A, Feldman HA, McKinlay JB, Longcope C (1991) Age, disease, and changing sex hormone levels in middle-aged men: results of the Massachusetts male aging study. J Clin Endocrinol Metab 73(5):1016–1025
Gross PA, Hermogenes AW, Sacks HS, Lau J, Levandowski RA (1995) The efficacy of influenza vaccine in elderly persons: a meta-analysis and review of the literature. Ann Intern Med 123:518–527
Greendale GA, Unger JB, Rowe JW, Seeman TE (1999) The relation between cortisol excretion and fractures in healthy older people. J Am Geriatr Soc 47(7):799–803
Grimley Evans J (2000) Ageing and medicine. J Intern Med 247:159–167
Grove GL, Kligman AM (1983) Age-associated changes in human epidermal cell renewal. J Gerontol 38:137–142
Graugaard-Jensen C, Schmidt F, Thomsen HF, Djurhuus JC (2008) Normal voiding patterns assessed by means of a frequency–volume chart. Scand J Urol Nephrol 42:269–273
Grubeck-Loebenstein B (1997) Changes in the aging immune system. Biologicals 25(2):205–208
Guha Mazumder DN, Chakraborty AK, Ghose A, Gupta JD, Chakraborty DP, Dey SB, Chattopadhyay N (1988) Chronic arsenic toxicity from drinking tubewell water in rural West Bengal. Bull WHO 66(4):499–506

Gurnell EM, Chatterjee VK (2001) Dehydroepiandrosterone replacement therapy. Eur J Endocrinol 145(2):103–106

Hak AE, Witteman JC, de Jong FH, Geerlings MI, Hofman A, Pols HA (2002) Low levels of endogenous androgens increase the risk of atherosclerosis in elderly men: the Rotterdam study. J Clin Endocrinol Metab 87(8):3632–3639

Hamilton HJ, Gallagher PF, O'Mahony D (2009) Inappropriate prescribing and adverse drug events in older people. BMC Geriatr 9:5. doi: 10.1186/1471-2318-9-5

Hanberg A, Hakansson H, Ahlborg UG (1989) "ED50" for TCDD-induction of body weight gain, liver enlargement, and thymic atrophy in Hartley guinea pigs, Sprague-Dawley rats, C57BL/6 mice, and golden Syrian hamsters. Chemosphere 9:813–816

Hanlon JT, Schmader KE, Koronkowski MJ, Weinberger M, Landsman PB, Samsa GP, Lewis IK (1997) Adverse drug events in high risk older outpatients. J Am Geriatr Soc 45:945–948

Hannan MT, Felson DT, Anderson JJ (1992) Bone mineral density in elderly men and women: results from the Framingham osteoporosis study. J Bone Miner Res 7:547–553

Hardman JG, Limbird LE (2001) Goodman & Gilman's the pharmacological basis of therapeutics, 10th edn. McGraw-Hill Medical Publishing Division, New York, NY

Harman SM, Metter EJ, Tobin JD, Pearson J, Blackman MR (2001) Longitudinal effects of aging on serum total and free testosterone levels in healthy men. Baltimore longitudinal study of aging. J Clin Endocrinol Metab 86(2):724–731

Harman SM, Tsitouras PD (1980) Reproductive hormones in aging men. I. Measurement of sex steroids, basal luteinizing hormone, and Leydig cell response to human chorionic gonadotropin. J Clin Endocrinol Metab 51(1):35–40

Harman D (1981) The aging process. Proc Natl Acad Sci USA 78(11):7124–7128

Hawkley LC, Cacioppo JT (2004) Stress and the aging immune system. Brain Behav Immun 18:114–119

Hayes JR, Condie LW Jr, Egle JL Jr, Borzelleca JF (1987) The acute and subchronic toxicity in rats of *trans*-1,2-dichloroethylene in drinking water. J Am Coll Toxicol 6:471–478

Hazzard WR, Blass JP, Ettinger WH, Jr, Halter JB, Ouslander JG (1999) Principles of geriatric medicine and gerontology, 4th edn. McGraw-Hill, Health Professions Division, New York, NY, pp 104–116

Heaton PC, Fenwick SR, Brewer DE (2007) Association between tetracycline or doxycycline and hepatotoxicity: a population I case–control study. J Clin Pharm Ther 32(5):483–487

Helzer EP, Cauley JA, Pratt SR, Wisniewski SR, Zmuda JM, Talbott EO, de Rekeneire N, Harris TB, Rubin SM, Simonsick EM, Tylavsky FA, Newman Anne B (2005) Race and sex differences in age-related hearing loss: the health, aging and body composition study. J Am Geriatr Soc 53(12):2119–2127

Henck JW, Quast JF, Rampy LW (1979) A comparison of four mouse strains exposed to subchronically inhaled vinylidene chloride (VDC). Toxicology Research Laboratory, Health and Environmental Science, Dow Chemical USA, Midland, MI

Hernandez-Zavala A, Del Razo LM, Aguilar C, Garcia-Vargas GG, Borja VH, Cebrian ME (1998) Alteration in bilirubin excretion in individuals chronically exposed to arsenic in Mexico. Toxicol Lett 99:79–84

Hodgson E, Rose RL (2007) Human metabolic interactions of environmental chemicals. J Biochem Mol Toxicol 21(4):182–186

Holick MF, Matsuoka LY, Wortsman J (1989) Age, vitamin D, and solar ultraviolet. Lancet 2: 1104–1105

Holmes KL, Schnitzlein CT, Perkins EH, Tew JC (1984) The effect of age on antigen retention in lymphoid follicles and collagenous tissue of mice. Mech Ageing Dev 25:243–249

Holsapple MP, Dooley RK, McNerney PJ, McCay JA (1986) Direct suppression of antibody responses by chlorinated dibenzodioxins in cultured spleen cells from (C57BL/6 × C3H)F1 and DBA/2 mice. Immunopharmacology 12:175–186

Homma Y, Imajo C, Takahashi S, Kawabe K, Aso Y (1994) Urinary symptoms and urodynamics in a normal elderly population. Scand J Urol Nephrol Suppl 157:27–30

Horan MA, Ashcroft GS (1997) Ageing, defence mechanisms and the immune system. Age Aging 26(S4):15–19

Hui-Ling L, Chen J-T, Ma T-C, Chang Y-S (2008) Analysis of drug–drug interactions (DDIs) in nursing homes in Central Taiwan. Arch Gerontol Geriatr 47:99–107

Hultman P, Enestrom S (1992) Dose–response studies in murine mercury-induced autoimmunity and immune-complex disease. Toxicol Appl Pharmacol 113(2):199–208

Hunt CM, Westerkarn SR, Stave M, Wilson J (1992a) Hepatic cytochrome p-4503A (CYP3A) activity in the elderly. Mech Ageing Dev 64:189–199

Hunt CM, Westerkam WR, Stave GM (1992b) Effect of age and gender on the activity of human hepatic CYP3A. Biochem Pharmacol 44:275–283

Hutterer F, Schaffner F, Klion FM, Popper H (1968) Hypertrophic, hypoactive smooth endoplasmic reticulum: a sensitive indicator of hepatotoxicity exemplified by dieldrin. Science 161(3845):1017–1019

Institoris L, Siroke O, Desi I, Undeger U (1999) Immunotoxicological examination of repeated dose combined exposure by dimethoate and two heavy metals in rats. Hum Exp Toxicol 18(2):88–94

IOM (1997) Institute of Medicine. DRI Dietary Reference Intakes for calcium, phosphorus, magnesium, vitamin D and fluoride. Standing Committee on the scientific evaluation of dietary reference intakes. Food and Nutrition Board, Institute of Medicine, National Academies Press, Washington, DC

Ishimaru T, Pages L, Horton R (1977) Altered metabolism of androgens in elderly men with benign prostatic hyperplasia. J Clin Endocrinol Metab 45(4):695–701

Jaeschke H, Gores GJ, Cederbaum AI, Hinson JA, Pessayre D, Lemasters JJ (2002) Mechanisms of hepatotoxicity. Toxicol Sci 65:166–176

Jernigan TL, Archibald SL, Fennema-Notestine C, Gamsi AC, Stout JC, Bonner J, Hesselink J (2001) Effects of age on tissues and regions of the cerebrum and cerebellum. Neurobiol Aging 22:581–594

Johnson SR (1998) Menopause and hormone replacement therapy. Med Clin North Am 82(2): 297–320

Jones A, Morton I, Hobson L, Evans GS, Everard ML (2006) Differentiation and immune function of human dendritic cells following infection by respiratory syncytial virus. Clin Exp Immunol 143(3):513–522

Jones G, Nguyen T, Sambrook P, Kelly PJ, Eisman JA (1994) Progressive loss of bone in the femoral neck in elderly people: longitudinal findings from the Dubbo osteoporosis epidemiology study. BMJ 309:691–695

Jonker D, Woutersen RA, van Bladeren PJ, Til HP, Feron VJ (1993) Subacute (4-week) oral toxicity of a combination of four nephrotoxicants in rats compared with the toxicity of the individual compounds. Food Chem Toxicol 31(2):125–136

Jourdan M, Vaubourdolle M, Cynober L, Aussel C (2004) Effect of aging on liver functions – an experimental study in a perfused rat liver model. Exp Gerontol 39(9):1341–1446

Junius-Walker U, Theile G, Hummers-Pradier E (2006) Prevalence and predictors of polypharmacy among older primary care patients in Germany. Fam Pract. doi:10,1903/fampra/cml067

Juurlink DN, Mamdani M, Kopp A, Laupacis A, Redeimeier DA (2003) Drug–drug interactions among elderly patients hospitalized for drug toxicity. JAMA 289(13):1599–1600

Jyrkka J, Enlund H, Korhonen MJ, Sulkava R, Hartikainen S (2009) Patterns of drug use and factors associated with polypharmacy and excessive polypharmacy in elderly persons. Drugs Aging 26(6):493–503

Kalaria RN, Mitchell MJ, Harik SI (1987) Correlation of 1-methyl-4-phenyl-1,2,3,6-tetrahydropyridine neurotoxicity with blood–brain barrier monoamine oxidase activity. Proc Natl Acad Sci USA 84:3521–3525

Kamijo Y, Soma K, Asari Y, Ohwada T (1998) Survival after massive arsenic poisoning self-treated by high fluid intake. Clin Toxicol 36(1–2):27–29

Kampmann JP, Sinding J, Moller-Jorgensen I (1975) Effect of age on liver function. Geriatrics 30:19–95
Kang I, Hong MS, Nolasco H, Park SH, Dan JM, Choi JY, Craft J (2004) Age-associated change in the frequency of memory CD4+ T cell responses to influenza vaccine. J Immunol 173:673–681
Kaplan C, Pasternack B, Shah H, Gallo G (1975) Age-related incidence of sclerotic glomeruli in human kidneys. Am J Pathol 80:227–234
Kaplowitz N, Aw TY, Simon FR, Stolz A (1986) Drug-induced hepatotoxicity. Ann Inter Med 104(6):5A6
Kappel B, Olsen S (1980) Cortical interstitial tissue and sclerosed glomeruli in the normal human kidney, related to age and sex: a quantitative study. Virchows Arch A Pathol Anat Histol 387(3):271–277
Kareskay L, Tjernlunel UM, Forsum U, Peterson PA (1977) Epidermal Langerhans cells express Ia antigens. Nature 268:248–250
Kasiske B (1987) Relationship between vascular disease and age-associated changes in the human kidney. Kidney Int 31:1153–1159
Kassianides C, Nussenblatt R, Palestine AG, Mellow SD, Hoofnagle JH (1990) Liver injury from cyclosporine A. Digest Dis Sci 35(6):693–697
Kato N, Yoshida A (1980) Effect of dietary PCB on hepatic cholesterogenesis in rats. Nutr Rep Int 21:107–112
Kelly RI, Pearse R, Bull RH, Leveque JL, de Rigal J, Mortimer PS (1995) The effects of aging on the cutaneous microvasculature. J Am Acad Dermatol 33:749–756
Kinirons MT, O'Mahony MS (2004) Drug metabolism and aging. Br J Clin Pharmacol 57(5): 540–544
Kipen HM, Cody RP, Goldstein BD (1989) Use of longitudinal analysis of peripheral blood counts to validate historical reconstructions of benzene exposure. Environ Health Perspect 82:199–206
Kirschmann DA, Murasko DM (1992) Splenic and inguinal lymph node T-cells of aged mice respond differently to polyclonal and antigen-specific stimuli. Cell Immunol 139:426–437
Klaassen CD (2001) Casarett & Doull's toxicology, the basic science of poisons. McGraw-Hill, New York, NY, pp 417–438
Kline GH, Hayden TA, Klinman NR (1999) B cell maintenance in aged mice reflects both increased B cell longevity and decreased B cell generation. J Immunol 162:3342–3349
Kochi S, Takanaga H, Matsuo H, Ohtani H, Naito M, Tsuruo T, Sawada Y (2000) Induction of apoptosis in mouse brain capillary endothelial cells by cyclosporin A and Tacrolimus. Life Sci 66:2255–2260
Koller LD (1977) Enhanced polychlorinated biphenyl lesions in Moloney leukemia virus-infected mice. Clin Toxicol 11:107–116
Kotsonis FN, Klaassen CD (1978) The relationship of metallothionein to the toxicity of cadmium after prolonged administration to rats. Toxicol Appl Pharmacol 46:39–54
Kowalski TE, Falestiny M, Furthe E, Malet PF (1994) Vitamin A hepatotoxicity: a cautionary note regarding 25,000 IU supplements. Am J Med 97(6):523–528
Kradjan WA, Driesner NK, Abuan TH, Emmick G, Schoene RB (1992) Effect of age on bronchodilator response. Chest 101:1545–1551
Kronenberg RS, Drage CW (1973) Attenuation of the ventilatory and heart rate responses to hypoxia and hypercapnia with aging in normal men. J Clin Invest 53:1812–1819
Kreutler PA, Varbanov V, Goodman W, Olaya G, Stanbury JB (1978) Interactions of protein deficiency, cyanide, and thiocyanate on thyroid function in neonatal and adult rats. Am J Clin Nutr 31:282–289
Krzystyniak K, Hugo P, Flipo D, Fournier M (1985) Increased susceptibility to mouse hepatitis virus 3 of peritoneal macrophages exposed to dieldrin. Toxicol Appl Pharmacol 80:397–408
Kumar V, Abbas AK, Fausto N (eds) (2005) Robins and Cotran pathologic basis of disease, 7th edn. Elsevier, Philadelphia, PA, p 706
Labat-Robert J, Kern P, Robert L (1992) Biomarkers of connective tissue aging: biosynthesis of fibronectin, collagen type II and elastase. Ann NY Acad Sci 673:116–122

Lakatta EG (1993) Cardiovascular regulatory mechanisms in advanced age. Physiol Rev 73:413
Lam CSP, Borlaug BA, Kane GC, Enders FT, Rodeheffer RJ, Redfield MM (2009) Age-associated increases in pulmonary artery systolic pressure in the general population. Circulation 119:2663–2670
Lange P, Schnohr P (2007) The relationship between facial wrinkling and airway obstruction. Int J Dermatol 33(2):123–126
Leape LL, Begrnnan TA, Laird N (1991) The nature of adverse events in hospitalized patients: results of the Harvard Medical Practice Study II. N Engl J Med 324:377–384
Lee CC, Bhandari JC, Winston JM, House WB (1977) Inhalation toxicity of vinyl chloride and vinylidene chloride. Environ Health Perspect 21:25–32
Leitolf H, Behrends J, Braband G (2002) The thyroid axis in aging. In: Chadwick DJ, Goode JA (eds) Novartis foundation symposium 242 – endocrine facets of aging. John Wiley and Sons, Chichester, UK ISBN 9780471486367
Lemaoult J, Szabo P, Weksler ME (1997) Effect of age on humoral immunity, selection of the B-cell repertoire and B-cell development. Immunol Rev 160:115–126
Le Page C, Noirez P, Courty J, Riou B, Swynghedauw B, Besse S (2009) Exercise training improves functional post-ischemic recovery in senescent heart. Exp Gerontol 44:177–182
Lesser PB, Vietti MM, Clark WD (1985) Lethal enhancement of therapeutic doses of acetaminophen by alcohol. Digest Dis Sci 32(1):193–195
Leuenberger N, Pradervand S, Wahli W (2009) Sumolyated PPARα mediates sex-specific gene repression and protects the liver from estrogen-induced toxicity in mice. J Clin Invest. doi: 10.1172/JCI39019
Levin-Scherz JK, Patrick JD, Weber FH, Garabeedian C Jr (1987) Acute arsenic ingestion. Ann Emerg Med 16(6):702–704
Lewis J, Zimmerman H (1999) Drug- and chemical-induced cholestasis. Clin Liver Dis 3(3): 433–464
Li L, Mac-Mary S, Marsaut D, Sainthiller JM, Nouveau S, Gharbi J, de Lacharriere O, Humbert P (2006) Age-related changes in skin topography and microcirculation. Arch Dermatol Res 297:412–416
Li RW, Edwards MH, Brown B (2001) Variation in vernier evoked cortical potential with age. Invest Ophthalmol Vis Sci 42:1119–1124
Li GL , Yin SN, Watanabe T, Nakatsuka H, Kasahara M, Abe H, Ikeda M (1986) Benzene-specific increase in leukocyte alkaline phosphatase activity in rats exposed to vapors of various organic solvents. J Toxicol Environ Health 19:581–589
Licht H, Seeff LB, Zimmerman HJ (1980) Apparent potentiation of acetaminophen hepatotoxicity by alcohol. Ann Inter Med 92(4):511
Lieber CS, DeCarli LM (1976) Animal models of ethanol dependence and liver injury in rats and baboons. Fed Proc 35:1232–1236
Lieber CS, Jones DP, DeCarli LM (1965) Effects of prolonged ethanol intake: production of fatty liver despite adequate diets. J Clin Invest 44:1009–1021
Lin W-Q, White, KL (1993) Production of complement component C3 in vivo following 2,3,7, 8-tetrachlorodibenzo-*p*-dioxin exposure. J Toxicol Environ Health 39:273–285
Lindeman RD (1993) Renal physiology and pathophysiology of aging. Contrib Nephrol 105: 1–12
Lindeman RD, Preuss HG (1994) Renal physiology and pathophysiology of aging. Geriatr Nephrol Urol 4:113–120
Lindeman RD, Tobin JD, Shock NW (1985) Longitudinal studies on the rate of decline in renal function with age. J Am Geriatr Soc 33:278–285
Linjakumpu, T, Hartikainen S, Klaukka T, Veijola J, Kivela S-L, Isoaho R (2002) Use of medications and polypharmacy are increasing among the elderly. J Clin Epidemiol 55(8): 809–817
Linton PJ, Haynes L, Klinman NR, Swain SL (1996) Antigen-independent changes in naïve CD4 T cells with aging. J Exp Med 184:1891–1900

Lipschitz DA, Udupa KB, Milton KY, Thompson CO (1984) Effect of age on hematopoiesis in man. Blood 63:502–509
Litterst CL, Rarber TM, Baker AM, Van Loon EJ (1972) Effects of polychlorinated biphenyls on hepatic microsomal enzymes in the rat. Toxicol Appl Pharmacol 23:112–122
Longcope C, Jaffee W, Griffing G (1980) Metabolic clearance rates of androgens and oestrogens in aging women. Maturity 2(4):283–290
Loose LD, Silkworth JB, Charbonneau T, Blumenstock F (1981) Environmental chemical-induced macrophage dysfunction. Environ Health Perspect 39:79–92
Lopez-Novoa JM (2008) The mechanisms of age-associated glomerular sclerosis. In: Machas JF, Cameron JS, Oreopoulos DG (eds) The aging kidney in health and disease. Springer, New York, NY, pp 113–126
Lord JM, Butcher S, Killampali V, Lascelles D, Salmon M (2001) Neutrophil ageing and immunosenescence. Mech Ageing Dev 122:1521–1535
Macefield G, Gandevia SC, Burke D (1989) Conduction velocities of muscle and cutaneous afferents in the upper and lower limbs of human subjects. Brain 112(6):1519–1532
Mahler DA, Rosiella RA, Loke J (1986) The aging ling. Geriatr Clin North Am 2:215–225
Makrantonaki E, Zouboulis CC (2007) The skin as a mirror of the aging process in the human organism – State of the art and results of the aging research in the German National Genome Research Network 2 (NGFN-2). Exper Gerontol 42(9):879–886
Malave I, de Ruffino DT (1984) Altered immune response during cadmium administration in mice. Toxicol Appl Pharmacol 74:46–56
Mallya KB, Mendis T, Guberman A (1986) Bilateral facial paralysis following ethylene glycol ingestion. Can J Neurol Sci 13(4):340–341
Maltoni C, Lefemine P, Cotti G, Patella V (eds) (1985) Experimental research on vinylidine chloride carcinogenesis: archives of research on industrial carcinogenesis, vol 3. Princeton Scientific, Princeton, NJ, p 229
Marchesini G, Bua V, Brunori A, Bianchi G, Ppisi P, Fabbri A, Zoli M, Pisi E (1988) Galactose elimination capacity and liver volume in ageing man. Hepatology 8:1079–1083
Mariotti S, Franceschi C, Cossarizza A, Pinchera A (1995) The aging thyroid. Endocr Rev 16(6):686–715
Marley SB, Lewis JL, Davidson RJ, Roberts IAG, Dokal I, Goldman JM, Gordon MY (1999) Evidence for a continuous decline in haemopoietic cell function from birth: application to evaluating bone marrow failure in children. Br J Haematol 106:162–166
Massey LK. (1998) Caffeine and the elderly. Drug Aging 13:45–50
Massey LK, Whiting SJ (1993) Caffeine, urinary calcium, calcium metabolism and bone. J Nutr 123:1611–1614
McCauley PT, Robinson M, Condie LW, Parvell M (1990) The effect of subacute and subchronic oral exposure to *cis*-1,2-dichloroethylene in rats. Environmental Protection Agency, Health Effects Research Laboratory, Cincinnati, OH, and Air Force Aerospace Medical Research Laboratory, Wright-Patterson AFB, OH
McConnel EE, Moore JA, Dalgard DW (1978) Toxicity of 2,3,7,8-tetrachlorodibenzo-*p*-dioxin (TCDD) in rhesus monkeys (*Macaca mulatta*) following a single oral dose. Toxicol Appl Pharmacol 43:175–187
McCullough JL, Kelly KM (2006) Prevention and treatment of skin aging. Ann NY Acad Sci 1067:323–331
McKenna MJ, Zempel JA, Madrid EO, Gehring PJ (1978) The pharmacokinetics of [14C] vinylidine chloride in rats following inhalation exposure. Toxicol Appl Pharmacol 45:599–610
McLean AJ, Le Couteur DG (2004) Aging biology and geriatric clinical pharmacology. Pharmacol Rev 56:163–184
Mege JL, Capo C, Michel B, Gastaut JL, Bongrand P (1988) Phagocytic cell function in aged subjects. Neurobiol Aging 9:217–220
Meier-Ruge W, Ulrich J, Bruhlmann N, Meier E (1992) Age-related white matter atrophy in the human brain. Ann NY Acad Sci 673:260–269

Milberg JA, Davis DR, Steinberg KP, Hudson LD (1995) Improved survival of patients with acute respiratory distress syndrome (ARDS): 1983–1993. JAMA 273:306–309
Milla P (2002) Children are not just little adults. Clin Nutr 21(Suppl 1):1370139
Miller RA (1996) The aging immune system: primer and prospectus. Science 273:70–74
Miller C, Kelsoe G, Han S (1994) Lack of B7-2 expression in the germinal centers of aged mice. Aging Immunol Infect Dis 5:249–257
Minuk GY, Kelly JK, Hwang W-S (1987) Vitamin A hepatotoxicity in multiple family members. Hepatology 8(2):272–275
Mittman C, Edelman NH, Norris AH, Shuck NW (1965) Relationship between chest wall and pulmonary compliance and age. J Appl Physiol 20:1211–1216
Mocchegiani E, Malavolta M (2004) NK and NKT cell functions in immunosenescence. Aging Cell 2004:177–184
Monette J, Gurwitz JH, Avorn J (1995) Epidemiology of adverse drug events in the nursing home setting. Drugs Aging 7(3):203–211
Montagna W, Carlisle K (1979) Structural damages in aging human skin. J Invest Dermatol 73:74
Mooradian AD (1992) Geriatric Neuroendocrinology. In: Nemeroff CB( ed) Neuroendocrinology. CRC Press, Boca Raton, FL, pp 541–562
Moore FA, Sauaia A, Moore EE, Haenel JB, Burch JM, Lezotte DC (1996) Postinjury multiple organ failure: bimodal phenomenon. J Trauma 0:5011–510
Morely JE, Reece SS (1989) Clinical implications of the aging heart. Am J Med 86:77–86
Moragas A, Castells C, Sans M (1993) Mathematical morphologic analysis of aging-related epidermal changes. Anal Quant Cytol Histol 15:75–82
Morely JE, Kaiser FE, Merry HMI, Patrick P, Morely PM, Stauber PM, Vellas B, Baumgartner RN, Garry PJ (1997) Longitudinal changes in testosterone, luteinizing hormone, and follicle-stimulating hormone in healthy older men. Metabolism 46(4):410–413
Morris JS, Schmidt M, Newman S, Schever PG, Sherlock S (1974) Arsenic and noncirrhotic portal hypertension. Gastroenterology 66(1):86–94
Moscicki EK, Elkins HM, McNamara PM (1985) Hearing loss in the elderly: an epidemiologic study of the Framingham Heart Study Cohort. Ear Hear 6(4):184–190
Muravchick S (1996) Anesthesia for the geriatric patient. In: Barfash PG (ed) Clinical anesthesia. Lippincott-Raven, Philadelphia, PA, pp 1131–1133
Mulligen T, Iranmanesh A, Johnson ML, Straume MJD (1997) Aging alters feed-forward and feedback linkages between LH and testosterone in healthy men. Am J Physiol 273 (4. pt. 2):R1407–1413
Murakami S, Nasu M, Fukayama H, Krishnam L, Sugawara M (2004) Propranolol has direct antithyroid activity: Inhibition of iodide transport in cultured thyroid follicles. Cell Biochem Funct 11(3):159–165
Murasko DM, Dernstein ED, Gardner EM, Gross P, Munk G, Dran S, Abrutyn E (2002) Role of humoral and cell-mediated immunity in protection from influenza 33 disease after immunization of healthy elderly. Exp Gerontol 37(2–3):427–439
Murray MD, Brater DC (1993) Renal toxicity of the nonsteroidal anti-inflammatory drugs. Ann Rev Pharmacol Toxicol 32:435–465
Murray FJ, Smith FA, Nitschke KD, Humiston CG, Kociba RJ, Schwetz BA (1979) Three-generation reproduction study of rats given 2,3,7,8-tetrachlorodibenzo-*p*-dioxin (TCDD) in the diet. Toxicol Appl Pharmacol 50:241–251
Nagahiro S, Yamamoto YL, Diksic M, Mitsuka S, Sugimoto S, Feindel W (1991) Neurotoxicity after intracarotid 1,3-bis(2-chloroethyl)-1-nitrosourea administration in the rat: hemodynamic changes studied by double-tracer autoradiography. Neurosurgery 29:19–25
Nagaratnam N, Ip J, Bou-Haidar P (2005) The vestibular dysfunction and anxiety disorder interface: a descriptive study with special reference to the elderly. Arch Geront Geriatr 40:253–264
NAS (2005) National Academies of Science. Health implications of perchlorate ingestion. Board on environmental studies and toxicology (BEST). Appendix D – Sensitivity of perchlorate

induced iodide uptake inhibition to serum iodide concentrations. The National Academies Press, pp 216–220
Naylor K, Li G, Vallejo AN, Won-Woo L, Koetz K, Bryl E, Witkowski J, Fulbright J, Weyand CM, Goronzy JJ (2005) The influence of age on T cell generation and TCR diversity. J Immunol 174:7446–7452
Neugarten J, Gallo G, Silbiger S, Kasiske B (1999) Glomerulosclerosis in aging humans is not influenced by gender. Am J Kidney Dis 34(5):884–888
Ng Y, Goldspink DF, Burniston JG, Clark WA, Colyer J, Tan L-B (2002) Characterization of isoprenaline myotoxicity on slow-twitch skeletal versus cardiac muscle. Int J Cardiol 86: 299–309
Nichols WW, O'Rourke MF (1998) Aging. In: Nichols WWW, O'Rourke MF (eds) McDonald's blood flow in arteries: theoretical, experimental and clinical principles. Oxford University Press, New York, NY, pp 355–358
Nordin BEC, Need AG, Morris HA, O'Loughlin PD, Horowitz M (2004) Effect of age on calcium absorption in post-menopausal women. Am J Clin Nutr 80:998–1002
NRC (1987) Work, aging, and vision. Report of a Conference. National Research Council, National Academy Press, Washington, DC
NTP (1993) Toxicology and carcinogenesis studies of mercuric chloride (CAS no. 7487-94-7) in F344/N rats and B6C3F1 mice (gavage studies). ational Toxicology Program, U.S. Department of Health and Human Services, Public Health Service, National Institutes of Health, Research Triangle Park, NC. NTP TR 408. NIH publication no. 91-3139
NTP (1985a) Toxicology and carcinogenesis studies of chlorobenzene (CAS No. 108-90-70) in F344/N rats and B6C3F mice (gavage studies). Technical report series no. 261. Research Triangle Park, NC. U.S. Department of Health and Human Services, Public Health Service, National Institutes of Health, National Toxicology Program. NIH publication no. 86-2517
NTP (1985b) Toxicology and carcinogenesis studies of 1,2-dichlorobenzene (*o*-dichlorobenzene) in F344/N rats and B6C3F1 mice (gavage studies). Research Triangle Park, NC. National Toxicology Program. NTP TR255. NIH publication no. 86-2511
Nunez C, Nishimoto N, Grland GL, Billips LG, Burrows PD, Kubagawa H, Cooper MD (1996) B cells are generated throughout life in humans. J Immunol 156:866–872
Oddie TH, Meade JH Jr, Fisher DA (1966) An analysis of published data on thyroxine turnover in human subjects. J Clin Endocrinol Metab 26(4):425–436
Ogawa T, Kitagawa M, Hirokawa K (2000) Age-related changes of human bone marrow: a histometric estimation of proliferative dells, apoptotic cells, T cells, B cells and macrophages. Mech Ageing Dev 117:57–68
Ohsawa M, Takahashi K, Otsuka F (1988) Induction of anti-nuclear antibodies in mice orally exposed to cadmium at low concentrations. Clin Exp Immunol 73:98–102
Oliver P, Bertrand L, Tubery M, Lauque D, Montastruc J-L, Layeyre-Mestre M (2009) Hospitalizations because of adverse drug reactions in elderly patients admitted through the emergency department. Drugs Aging 26(6):475–482
Olivetti G, Melissari M, Capasso JM, Anversa P (1991) Cardiomyopathy of the aging human heart. Myocyte loss and reactive cellular hypertrophy. Circ Res 68:1560–1568
Orentreich N, Brind JL, Rizer RL, Vogelman JH (1984) Age changes and sex differences in serum dehydroepiandrosterone sulfate concentrations throughout adulthood. J Clin Endocrinol Metab 59(3):551–555
Orwoll E, Nielson CM, Marshall LM, Lambert L, Holton KF, Hoffman AR, Barrett-Connor E, Shikany JM, Dam T, Cauley JA (2009) Vitamin D deficiency in older men. J Clin Endocrinol Metab 94:1214–1222
O'Shea J (1997) Jaks, STATs, cytokine signal review, transduction, and immunoregulation: are we there yet? Immunity 7:1–11
Oskvig RM (1999) Special problems in the elderly. Chest 115:158S–164S
Park SK, Prolla TA (2005) Lessons learned from gene expression profile studies of aging and caloric restriction. Ageing Res Rev 4:55–65

Parker BM, Cusack BJ, Vestal RE (1995) Pharmacokinetic optimisation of drug therapy in elderly patients. Drugs Aging 7:10–18

Partridge L, Mangel M (1999) Messages from mortality: the evolution of death rates in the old. Trends Ecol Evol 14(11):438–442

Pawelec G, Rehbein A, Haehnel K, Meri A. Adibzadeh M (1997) Human T cell clones as a model of immunosenescence. Immunol Rev 160:31–43

Pawelec G, Solana R, Remarque E, Mariani E (1998) Impact of aging on innate immunity. J Leuko Biol 6:703–712

PDR (Physicians' Desk Reference) (2007) Physicians' desk reference, 61st edn. Medical Economics, Montvale, NJ, p 3153

Pearce D, Bonnet D (2009) Ageing within the hematopoietic stem cell compartment. Mech Ageing Dev 130:54–57

Penumarthy L, Oehme FW (1975) Treatment of ethylene glycol toxicosis in cats. Am J Vet Res 36(2):209–212

Perl DP, Gajdusek DC, Garruto RM, Yanagihara RY, Gibbs CJ (1980) Intraneuronal aluminum encephalopathy. Acta Neuropathol 50:19–24

Perry HM III (1999) The endocrinology of aging. Clin Chem 45(8B):1369–1376

Pfefferbaum A, Mathalson DH, Sullivan EV, Rawles JM, Zipursky RB, Lim KO (1994) A quantitative magnetic resonance imaging study of changes in brain morphology from infancy to late adulthood. Arch Neurol 51(9):874–887

Pochi JS, Strauss JS, Downing DT (1979) Age-related changes in sebaceous gland activity. J Invest Dermatol 73:106–111

Poligano A, Tortorella C, Venezia A, Jirillo E, Antonaci S (1994) Age-associated changes of neutrophil responsiveness in a human healthy elderly population. Cytobios 80: 145–153

Pontoppidan H, Beecher HK (1960) Progressive loss of protective reflexes in the airway with advanced age. JAMA 2209–2213

Price SC, Ozalp S, Weaver R, Chescoe D, Mullervy J, Hinton RH (1988) The thyroid hyperactivity caused by hypolipodaemic compounds and polychlorinated biphenyls: the effect of coadministration in the liver and thyroid. Arch Toxicol Suppl 12:85–92

Prigge E (1978) Early signs of oral and inhalative cadmium uptake in rats. Arch Toxicol 40: 231–247

Priox J, Romonatxo M, Hayot M, Mucci P, Prefaut C (2000) Effect of aging on the ventilatory response and lactate kinetics during incremental exercise in man. Eur J Appl Physiol 81 (1–2):100–107

Proudlock FA, Shekhar H, Gottlob I (2004) Age-related changes in head and eye coordination. Neurobiol Aging 25:1377–1385

Qato DM, Alexander GC, Conti RM, Johnson M, Schumm P, Lindau ST (2008) Use of prescription and over-the-counter medications and dietary supplements among older adults in the United States. JAMA 300(24):2867–2878

Rabe JH, Mamelak AJ, McElgunn PJ, Morison WL, Sauder DN (2006) Photoaging: mechanisms and repair. J Am Acad Dermatol 55:1–19

Rapuri PB, Gallagher JC, Kinyamu HK, Ryschon KL (2001) Caffeine intake increases the rate of bone loss in elderly women and interacts with vitamin D receptor genotypes. Am J Clin Nutr 74:694–700

Ravelojaona V, Robert L, Robert A-M (2008) Effect of cellular aging on collagen biosynthesis II. Collagen synthesis and deposition by a human skin fibroblast strain over 25 passages. Arch Gerontol Geriatr 47:368–376

Raz N, Gunning FM, Head D, Dupuis, JH, McQuain J, Briggs SD, Thornton AE, Loken WJ, Acker JD (1997) Selective aging of the human cerebral cortex observed in vivo: differential vulnerability of the prefrontal gray matter. Cereb Cortex 7:268–282

Reenstra WR, Yaar M, Gilchrest BA (1996) Aging affects epidermal growth factor receptor phosphorylation and traffic kinetics. Exp Cell Res 227:252–255

Rego EM, Garcia AB, Viana SR, Falcao RP (1998) Age-related changes of lymphocyte subsets in normal bone marrow biopsies. Cytometry 34:22–29

Reidenberg MM (1982) Drug interactions and the elderly. J Am Geriatr Soc 30:567–568

Riachi NJ, Behmand RA, Harik SI (1991) Correlation of MPTP neurotoxicity in vivo with oxidation of MPTP by the brain and blood–brain barrier in vitro in five rat strains. Rain Res 555:19–24

Rieser C, Papesh C, Herold M, Bock G, Ramoner R, Klocker JH, Bartsch G, Thurnher M (1998) Differential deactivation of human dendritic cells by endotoxin desensitization: role of tumor necrosis factor-alpha and prostaglandin E2. Blood 91:3112–3117

Riggs BL, Melton LJI (1986) Involutional osteoporosis. N Engl J Med 314(26):1676–1686

Risher JF, De Rosa CT (1997) The precision and limitations of health guidance values. Hum Ecol Risk Assess 3(5):681–700

Roberts-Thompson IC, Yuonscaiyud V, Whittingham S (1974) Aging, immune response and mortality. Lancet II:368–370

Robinson SN, Shah R, Wong BA, Wong VA, Farris GM (1997) Immunotoxicological effects of benzene inhalation in male Sprague-Dawley rats. Toxicology 119(3):227–237

Rodriguez GL, Larbarthe DR, Huang B, Lopez-Gomez J (1994) Rise in blood pressure with age: new evidence of population differences. Hypertension 24:779–785

Romero IA, Ray DE, Chan MW, Abbott NJ (1996) An in vitro study of *m*-dinitrobenzene toxicity on the cellular components of the blood–brain barrier, astrocytes and endothelial cells. Toxicol Appl Pharmacol 139:94–101

Rothschild JM, Bates DW, Leape LL (2000) Preventable medical injuries in older patients. Arch Intern Med 160:2717–2728

Rowden G, Lewis MG, Sullivan AK (1977) Ia antigen expression on human epidermal Langerhans cells. Nature 268:247–248

Ruby CM, Hanlon JT, Fillenbaum GG, Pieper CF, Branch LG, Bump RC (2005) Medication use and control of urination among community-dwelling older adults. J Aging Health 17(5): 661–674

Ruscetti FW, Keller JR, Longo DL (1998) Chapter 105: Hematopoiesis. In Fauci AS, Braunwald E, Isselbacher KJ, Wilson JD, Martin JB, Kasper DL (eds) Harrison's principles of internal medicine, 14th edn. McGraw-Hill Medical, New York, NY, p 624

Russel-Aulet M, Jaffe CA, Mott-Friberg R, Barkan AL (1999) In vivo semiquantification of hypothalamic growth-hormone-releasing hormone (GHRH) output in humans: evidence for relative GNRH deficiency in aging. J Clin Endocrinol Metab 84(10):3490–3497

Sabbieti MG, Agas D, Sontoni G, Materazzi S, Menghi G, Marchetti L (2009) Involvement of p53 in phthalate effects on mouse and rat osteoblasts. J Cell Biochem 107:316–327

Salat DH, Tuch DS, Greve DN, vaan derr Kouwe AJ, Hevelone ND, Zelata AK, Rosen BR, Fischl B, Corkin S, Rosas HD, Dale AM (2005) Age-related alterations in white matter microstructure measured by diffusion tensor imaging. Neurobiol Aging 26:1215–1227

Sampson HW (2002) Alcohol and other factors affecting osteoporosis risk in women. Alcohol Res Health 26(4):292–298

Sauder DN (1986) Effect of age on epidermal immune function. Dermatol Clin 4:447

Saurwein-Teissl M, Lung TL, Marx F, Gschosser C, Asch E, Blasko I, Parson W, Bock G, Schonitzer D, Trannoy E, Grubeck-Loebenstein B (2002) Lack of antibody production following immunization in old age: association with CD8(+)CD28(−) T cell clonal expansions and an imbalance in the production of Th1 and Th2 cytokines. J Immunol 168(11):5893–5899

Schiffman SS (2007) Critical illness and changes in sensory perception. Proc Nutr Soc 66:331–345

Schiffman SS, Rogers MO, Zervakis J (2003) Loss of taste, smell, and other senses with age: effects of medication. In: Bales CW, Ritchie CS (ed) Handbook of clinical nutrition and aging. Humana Press, Totowa, NG, pp 211–289

Seeman TE, McEwen BS, Singer BH, Albert MS, Rowe JW (1997) Increase in urinary cortisol excretion and memory declines: MacArthur studies of successful aging. J Clin Endocrinol Metab 82(8):2458–2465

Serot JM, Christmann D, Dubost T, Couturier M (1997) Cerebrospinal fluid transthyretin: aging and late onset Alzheimer's disease. J Neurol Neurosurg Policy 63:506–508
Shah GN, Mooradian AD (1997) Age-related changes in the blood–brain barrier. Exper Gerontol 32(4, 5):501–519
Sharkey JR, Browne B, Ory MG, Wang S (2005) Patterns of therapeutic prescription medication category use among community-dwelling homebound older adults. Pharmacoepidemiol Drug Saf 14(10):715–723
Shiraha H, Gupta K, Drabik K, Wells A (2000) Aging fibroblasts present reduced epidermal growth factor (EGF) responsiveness due to preferential loss of EGF receptors. J Biol Chem 275(25):19343–19351
Shiwin C, Lin Y, Zhineng H, Xianzu, Zhaolu Y, Huidong X (1990) Cadmium exposure and health effects among residents in an irrigation area with ore dressing wastewater. Sci Total Environ 90:67–73
Sherman BM, West JH, Korenman SG (1976) The menopausal transition: analysis of LH, FSH, estradiol, and progesterone concentrations during menstrual cycles of older women. J Clin Endocrinol Metab 42(4):629–636
Siew S, Matta RK, Johnson M (1975) Investigation of "crystallosis" in ethylene glycol. Toxicity. Scan Electron Microsc 8:555–562
Silkworth JB, Cutler DS, Sack GG (1989) Immunotoxicity of 2,3,7,8-tetrachlorodibenzo-*p*-dioxin in a complex environmental mixture from the Love Canal. Fundam Appl Toxicol 12:303–312
Sinclair J, Jeffery E, Wrighton S, Kostrubsky V, Szakacs J, Wood S (1998) Alcohol-mediated increases in acetaminophen hepatotoxicity: role of CYP2E and CYP3A. Biochem Pharmacol 55(10):1557–1565
Slikker W, Vore M, Bailey JR, Meyers M, Montgomery C (1983) Hepatotoxic effects of estradiol-17 beta-D-glucuronide in the rat and monkey. Pharmacol Exp Ther 225(1):138–143
Smialowicz RJ, Andrews JE, Riddle MM, Rogers RR, Luebke RW, Copeland CB (1989) Evaluation of the immunotoxicity of low level PCB exposure in the rat. Toxicology 56:197–211
Smith JB, Fenske NA (1996) Cutaneous manifestations and consequences of smoking. J Am Acad Dermatol 34(5 Pt 1):717–732
Snyder Ca, Sellakumar AR, James DJ, Albert RE (1988) The carcinogenicity of discontinuous inhaled benzene exposures in CD-1 and C57BL/6 mice. Arch Toxicol 62:331–335
Snyder CA, Bernstein BD, Sellakumar A (1984) Evidence for hematotoxicity and tumorigenesis in rats exposed to 100 ppm benzene. Am J Ind Med 5:429–434
Snyder CA, Goldstein BD, Sellakumar AR, Bromberg I, Laskin S, Albert RE (1980) The inhalation toxicology of benzene: incidence of hematopoietic neoplasms and hematotoxicity in AKR/J and C57BL/6 J mice. Toxicol Appl Pharmacol 54:323–331
Snyder CA, Goldstein BD, Sellakamur A, Wolman S, Bromberg I, Erlichman MN, Laskin S (1978) Hematotoxicity of inhaled benzene to Sprague-Dawley rats and AKR mice at 300 ppm. J Toxicol Environ Health 4:605–618
Spearman ME, Leibman KG (1984) Effects of aging on hepatic and pulmonary glutathione *S*-transferase activities in male and females Fischer 344 rats. Biochem Pharmacol 33(8): 1309–1313
Speroff L, Glass RH, Kase NG (eds) (1999) Clinical gynecologic endocrinology and infertility, 6th edn. Lippincott Williams and Wilkins, Baltimore, MD
Steele AC, Kohli N, Mallipeddi P, Karram M (1999) Pharmacologic causes of female incontinence. Intern J Urogynocol 10:106–110
Steger M, Maczek C, Grubeckloebenstein B (1997) Peripheral blood dendritic cells reinduce proliferation in in vitro aged T-cell populations. Mech Ageing Dev 93:125–130
Steinmaus C, Miller MD, Howd R (2007) Impact of smoking and thiocyanate on perchlorate and thyroid hormone associations in the 2001–2002 National health and nutrition examination survey. Environ Health Perspect 1151(9):1333–1338
Stowe HD, Wilson M, Goyer RA (1972) Clinical and morphological effects of oral cadmium toxicity in rabbits. Arch Pathol 94:389–405

Stingl G, Katz SI, Clement L, Green I, Shevach EM (1978) Immunologic functions of Ia-bearing epidermal Langerhans cells. J Immunol 121:2005–2013

Stratton JR, Levy WC, Cerqueira MD, Schwartz RS, Abrass IB (1994) Cardiovascular responses to exercise. Effects of aging and exercise training in healthy men. Circulation 89: 1648–1655

Strazielle N, Ghersi-Egea JF (1999) Demonstration of a coupled metabolism-efflux process at the choroid plexus as a mechanism of brain protection toward xenobiotics. J Neurosci 19(15): 6275–6289

Street JC, Sharma RP (1975) Alteration of induced cellular and humoral immune responses by pesticides and chemicals of environmental concern: quantitative studies of immunosuppression by DDT, Aroclor 1254, carbaryl, carbofuran, and methyl parathion. Toxicol Appl Pharmacol 32:587–602

Swift ME, Burns AL, Gray KL, DiPietro LA (2001) Age-related alterations in the inflammatory response to dermal injury. J Invest Dermatol 117:1027–1035

Swift CG, Homeida M, Halliwell M, Roberts CJC (1984) Antipyrine disposition and liver size in elderly. Eur J Clin Pharmacol 14:149–151

Syntichaki P, Tavernarakis N (2006) Signaling pathways regulating protein synthesis during ageing. Exp Gerontol 41:1020–1025

Szakal AK, Taylor JK, Smith JP, Kosco MK, Burton GF, Tew JG (1988) Morphometry and kinetics of antigen transport and developing antigen retaining reticulum of follicular dendritic cells in lymph nodes of aging immune mice. Aging Immunol Infect Dis 1:7–22

Szumanska G, Gadamski R, Albrecht J (1993) Changes of the Na/K ATPase activity in the cerebral cortical microvessels of rat after single intraperitoneal administration of mercuric chloride: histochemical demonstration with light and electron microscopy. Acta Neuropathol 86: 65–70

Takafumi O, Kitagawa M, Hirokawa K (2000) Age-related changes of human bone marrow: a histometric estimation of proliferative cells, apoptotic cells, T cells, B cells and macrophages. Mech Ageing Dev 117:57–68

Tan L-B, Littler WA (1990) Measurement of cardiac reserve in cardiogenic shock: implications for prognosis and management. Br Heart J 64:121–128

Tauchi H, Tsubol L, Okutomi J (1971) Age changes in the human kidney of the different races. Gerontologia 17:87–97

Tavernarakis N, Driscoll M (2002) Caloric restriction and lifespan: a role for protein turnover? Mech Ageing Dev 123:215–229

Thomas PT, Hinsdill RD (1978) Effect of polychlorinated biphenyls on the immune responses of Rhesus monkeys and mice. Toxicol Appl Pharmacol 44:41–51

Tiefenbach B, Lange P (1980) Studies on the action of dimethoate on the immune system. Arch Toxicol Suppl 4:167–170

Timaffy M (1962) A comparative study of bone marrow function in young and old individuals. Gerontol Clin 4:13–18

Toft K, Olofsson T, Tunek A, Berlin M (1982) Toxic effects on mouse bone marrow caused by inhalation of benzene. Arch Toxicol 51:295–302

Tolep K, Higgins N, Muza S, Criner G, Kelsen SG (1995) Comparison of diaphragm strength between healthy adult elderly. Am J Resp Crit Care Med 152(2):677–682

Tolep K, Kelsen SG (1993) Effect of aging on respiratory skeletal muscles. Clin Chest Med 14(3):363–378

Tonn T, Esser C, Schneider EM, Steinmann-Steiner-Haldenstätt W, Gleichmann E (1996) Persistence of decreased T-helper cell function in industrial workers 20 years after exposure to 2,3,7,8-tetrachlorodibenzo-*p*-dioxin. Environ Health Perspect 104:422–426

Truelove J, Grant D, Mes J, Trypnonas H, Tryphonas L, Zawidzka ZZ (1982) Polychlorinated biphenyl toxicity in the pregnant cynomolgus monkey: a pilot study. Arch Environ Contam Toxicol 11:583–588

Tryphonas H, Hayward S, O'Grady L, Loo JCK, Arnold DL, Bryce F, Zawidzka ZZ (1989) Immunotoxicity studies of PCB (Aroclor 1254) in adult Rhesus (*Macaca mullata*) monkey – Preliminary report. Int J Immunopharmacol 11:199–206

Turnheim K (1998) Drug dosage in the elderly: is it rational? Drugs Aging 13:357–379

Umbreit TH, Patel D, Gallo MA (1985) Acute toxicity of TCDD contaminated soil from an industrial site. Chemosphere 14:945–947

Van CE, Leproult R, Kupfer DJ (1996) Effects of gender and age on the levels and circadian rhythmicity of plasma cortisol. J Clin Endocrinol Metab 81(7):2468–2473

Van Birgelen APJM, Van der Kolk, J, Fase KM, Bol I, Polger H, Brouwer A, Vandenberg M (1995) Subchronic dose–response study of 2,3,7,8-tetrachlorodibenzo-*p*-dioxin in female Sprague-Dawley rats. Toxicol Appl Pharmacol 132:1–13

Van Sande J, Deneubourg F, Beauwens R, Braekman JC, Daloze D, Dumont JE (1990) Inhibition of iodide transport in thyroid cells by dysidenin, a marine toxin, and some of its analogs. Mol Pharmacol 37(4):583–589

Varani J, Dame MK, Rittie L, Fligiel SEG, Kang S, Fisher GJ, Vorhees JJ (2006) Decreased collagen production in chronically aged skin. Am J Pathol 168(6):1861–1868

Vecchi AS, Sironi M, Canegrati MA, Recchia M, Garatini S (1983) Comparison of the immunosuppressive effects in mice of 2,3,7,8-tetrachlorodibenzo-*p*-dioxin and 2,3,7,8-tetrachlorodibenzofuran. In: Choudhary G, Keith LH, Rappe C (eds) Chlorinated dioxins and dibenzofurans in the total environment. Butterworth, Boston, MA, pp 397–402

Veldhuis JD, Iranmanesh A, Bowers CY (2005) Joint mechanisms of impaired growth-hormone pulse renewal in aging men. J Clin Endocrinol Metab 9(7):4177–4183

Vermeulen A (2001) Androgen replacement therapy in the aging male – a critical evaluation. J Clin Endocrinol Metab 86(6):2380–2390

Vermeulen A (1995) Dehydroepiandrosterone sulfate and aging. Ann NY Acad Sci 774:121–127

Vetricka V, Tlaskolova-Hogenova H, Pospisil M (1985) Impaired antigen-presenting function of macrophages form aged mice. Immunol Invest 14:105–114

Viluksela M, Stahl BU, Rozman KK (1994) Subchronic (13-week) toxicity of heptachlorodibenzo-*p*-dioxin in male Sprague-Dawley rats. Chemosphere 29:2381–2393

Vos JG, Moore JA, Zinkl JG (1973) Effect of 2,3,7,8-tetrachlorodibenzo-*p*-dioxin on the immune system of laboratory animals. Environ Health Perspect 5:149–162

Vos JG, de Roij T (1972) Immunosuppressive activity of a polychlorinated biphenyl preparation on the humoral immune response tin guinea pigs. Toxicol Appl Pharmacol 12:549–555

Vos JG, Van Dreil-Grootenhuis D (1972) PCBV-induced suppression of the humoral and cell-mediated immunity in guinea pigs. Sci Total Environ 1:289–302

Wagner W, Horn P, Bork S, Ho AD (2008) Aging of hematopoietic stem cells is regulated by the stem cell niche. Exp Gerontol 43:974–980

Waldmann TA (2006) The biology of interleukin-2 and interleukin-15: implications for cancer therapy and vaccine design. Nat Rev Immun 6(8):595–601

Waldmann TA, Tagaya Y (1999) The multifaceted regulation of interleukin-15 expression and the role of this cytokine in NK cell differentiation and host response to intracellular pathogens. Ann Rev Immunol 17:19–49

Walhovd KB, Fjell AM, Reinvang I, Lundervold A, Dale AM, Eilertsen DE, DE, Quinn BT, Salat D (2005) Effects of age on volumes of cortex, white matter, and subcortical structures. Neurobiol Aging 26:1261–1270

Walker J, Wynne H (1994) Review: the frequency and severity of adverse drug interactions in elderly people. Age Aging 23(3):255

Walker RM, Racz WJ, McElligott TF (1983) Scanning electron microscopic examination of acetaminophen-induced hepatotoxicity and congestion in mice. Am J Pathol 113: 321–330

Ward CO, Kuna RA, Snyder NK, Alsaker RD, Coate WB, Craig PH (1985) Subchronic inhalation toxicity of benzene in rats and mice. Am J Ind Med 7:457–473

Ware JH, Dockery DW, Louis TA, Xu X, Ferris BG Jr, Speizer FE (1990) Longitudinal and cross-sectional estimates of pulmonary function decline in never-smoking adults. Am J Epidemiol 132:685–700

Warfvinge K, Hansson H, Hultman P (1995) Systemic autoimmunity due to mercury vapor exposure in genetically susceptible mice: dose–response studies. Toxicol Appl Pharmacol 132:299–309

Weksler ME, Hutteroth TH (1974) Impaired lymphocyte function in aged humans. J Clin Invest 53:99–104

Wells MS, Nerland DE (1991) Hematotoxicity and concentration-dependent conjugation of phenol in mice following inhalation exposure to benzene. Toxicol Lett 56(1–2):159–166

Westerhof N, O'RourkeMF (1995) Haemodynamic basis for development of left ventricular failure in systolic hypertension and for it logical therapy. J Hyperten 13:943–952

Westphal JF, Vetter D, Brogard JM (1994) Hepatic side-effects of antibiotics. J Antimibrob Chemother 33:387–401

White KL Jr, Lysy HH, McCay JA, Anderson AC (1986) Modulation of serum complement levels following exposure to polychlorinated dibenzo-*p*-dioxins. Toxicol Appl Pharmacol 84:209–219

WHO (2005) Children are not little adults. World Health Organization. Children's Health and the Environment, WHO Training Package for the Health Sector. Geneva, Switzerland. Accessed 28/06/2005. www.who.int/ceh

Wicherts IS, van Schoor NM, Boeke AJP, Visser M, Deeg DJH, Smit J, Knol DL, Lips P (2007) Vitamin D status predicts physical performance and its decline in older persons. J Clin Endocrinol Metab 92(6):2058–2065

Wick G, Grubeck-Loebenstein B (1997) The aging immune system: primary and secondary alterations of immune reactivity in the elderly. Exp Gerontol 32(4–5):401–413

Wilcox SM, Himmelstein DU, Woolhandler S (1994) Inappropriate drug prescribing for the community-dwelling elderly. JAMA 272(4):292–296

Williamson J, Chopin JM (1980) Adverse reactions to prescribed drugs in the elderly: a multicentre investigation. Age Aging 9:73–80

Williams SG, Cooke GA, Wright DJ, Parsons WJ, Riley RL, Marshal P, Taqn L-B (2001) Peak exercise cardiac power outcome: a direct indicator of cardiac function strongly predictive of prognosis in chronic heart failure. Eur Heart J 22:1496–1503

Williams DA, Xu H, Cancelas JA (2006) Children are not little adults: just ask their hematopoietic stem cells. J Clin Invest 116(10):2593–2596

Wolf OT, Dziobek I, McHugh P, Sweat V, de Leon MJ, Javier E, Convit A (2005) Subjective memory complaints in aging are associated with elevated cortisol levels. Neurobiol Aging 26:1357–1363

Wolf MA, Rowe VK, McCollister DD, Hollingsworth RL, Oyen F (1956) Toxicological studies of certain alkylated benzenes and benzene. Arch Ind Health 14:387–398

Wolfson L (2001) Gait and balance dysfunction: a model of the interaction of age and disease. Neuroscientist 7(2):178–183

Woodhouse KW, Wynne HA (1988) Age-related changes in liver size and hepatic blood flow. The influence on drug metabolism in the elderly. Clin Pharmacokinet 15:287–294

Woodhouse KW, Mutch E, Williams FM, Rawlins MD, James OFW (1984) The effect of aging on pathways of drug metabolism in human liver. Age Aging 13:328–344

Worner TM, Lieber CS (1985) Perivenular fibrosis as precursor lesion of cirrhosis. JAMA 254:627–630

Wu AW (2000) Adverse drug events and near misses: who's counting? Am J Med 109(2):166–168

Wynne HA (2002) Age-related changes in liver, gall bladder and pancreas. Rev Clin Gerontol 12:12–20

Wynne HA, Cope LH, Mutch E, Rawlins MD (1989a) The effect of age upon liver volume and apparent blood flow in health man. Hepatology 9:297–301

Wynne HA, Cope LH, James OFW, Rawlins MD, Woodhouse KW (1989b) The effect of age and frailty upon acetanilide clearance in man. Age Aging 18(6):415–418

Wynne HA, Dope LH, Mutch E, Rawlins MD, Woodhouse KW, James OFW (1989c) The effect of age upon liver volume and apparent liver blood flow in healthy man. Hepatology 9(2):297–301

Xia X-L, Xi-Peng J, Pei-Lian L, Gu XQ, LaPorte RE, Tajima N (1995) Ascertainment corrected prevalence rate (ACPR) of leucopenia in workers exposed to benzene in small-scale industries calculated with capture-recapture methods. Biomed Environ Sci 8:30–34

Yaar M, Gilchrest BA (2001) Skin aging. Postulated mechanisms and consequent changes in structure and function. Geriatr Dermatol 17(4):617–630

Yamaguchi M, Ozaki K (1990) Aging affects cellular zinc and protein synthesis in the femoral diaphysis of rats. Res Exp Med 190:295–300

Yamamoto Y, Moore R, Hess HA, Guo GL, Gonzales FJ, Korach KS, Maronpot RR, Negishi M (2006) Estrogen receptor α mediates 17α-ethynylestradiol causing hepatotoxicity. J Biol Chem 281(24):16625–16631

Yueh B, Shapiro N, MacLean CH, Shekelle PG (2003) Screening and management of adult hearing loss in primary care: scientific review. JAMA 289(15):11976–1985

Yoon P, Keylock KT, Hartman ME, Freund GG, Woods JA (2004) Macrophage hypo-responsiveness to interferon-G in aged mice is associated with impaired signaling through Jak-STAT. Mech Ageing Dev 125:137–143

Zeeh J, Platt D (2002) The aging liver: structural and functional changes and their consequences for drug treatment in old age. Gerontology 48(3):121–127

Zhan C, Sangl J, Bierman AS, Miller MR, Friedman B, Wickizer SW, Meyer GS (2001) Potentially inappropriate medication use in the community-dwelling elderly: findings from the 1996 Medical Expenditure Panel Survey. JAMA 286(22):2823–2829

Zharhary D (1988) Age-related changes in the capability of the bone marrow to generate B cells. J Immunol 141:1863–1869

Zheng, W (2001) Neurotoxicology of the brain barrier system: new implications. Clin Toxicol 39(7):711–719

Zheng W, Blaner WS, Zhao Q (1999) Inhibition by Pb of production and secretion of transthyretin in the choroid plexus: its relationship to thyroxine transport at the blood–CSF barrier. Toxicol Appl Pharmacol 155:24–31

Zheng W, Shen H, Blaner SB, Zhao Q, Ren X, Graziano JH (1996) Chronic lead exposure alters transthyertin concentration in rat cerebrospinal fluid: the role of the choroid plexus. Toxicol Appl Pharmacol 139:445–450

Zimmerman HJ, Maddrey WC (1995) Acetaminophen (paracetamol) hepatotoxicity with regular intake of alcohol: analysis of instances of therapeutic misadventure. Hepatology 22(3):767–773

Zouboulis CC, Boschnakow A (2001) Chronological ageing and photoageing of the human sebaceous gland. Clin Exp Dermatol 26:600–607

# Index




D.M. Whitacre (ed.), *Reviews of Environmental Contamination and Toxicology*, Reviews of Environmental Contamination and Toxicology 207,
DOI 10.1007/978-1-4419-6406-9, © Springer Science+Business Media, LLC 2010

MIX
Papier aus verantwortungsvollen Quellen
Paper from responsible sources
FSC® C105338

If you have any concerns about our products,
you can contact us on
**ProductSafety@springernature.com**

In case Publisher is established outside the EU,
the EU authorized representative is:
**Springer Nature Customer Service Center GmbH**
**Europaplatz 3, 69115 Heidelberg, Germany**

Printed by Libri Plureos GmbH
in Hamburg, Germany